ENERGY SCIENCE, ENGINEERING AND TECHNOLOGY

W0079832

SYNGAS GENERATION FROM HYDROCARBONS AND OXYGENATES WITH STRUCTURED CATALYSTS

ENERGY SCIENCE, ENGINEERING AND TECHNOLOGY

Additional books in this series can be found on Nova's website under the Series tab.

Additional E-books in this series can be found on Nova's website under the E-book tab.

Energy Science, Engineering and Technology

Syngas Generation from Hydrocarbons and Oxygenates with Structured Catalysts

Vladislav Sadykov,
Ludmila Bobrova, Svetlana Pavlova,
Valentina Simagina, Lev Makarshin,
Valentin Parmon, Julian R. H. Ross,
Andre C. Van Veen
and
Claude Mirodatos

Nova Science Publishers, Inc.
New York

For permission to use material from this book please contact us:
Telephone 631-231-7269; Fax 631-231-8175
Web Site: http://www.novapublishers.com

NOTICE TO THE READER

The Publisher has taken reasonable care in the preparation of this book, but makes no expressed or implied warranty of any kind and assumes no responsibility for any errors or omissions. No liability is assumed for incidental or consequential damages in connection with or arising out of information contained in this book. The Publisher shall not be liable for any special, consequential, or exemplary damages resulting, in whole or in part, from the readers' use of, or reliance upon, this material.

Independent verification should be sought for any data, advice or recommendations contained in this book. In addition, no responsibility is assumed by the publisher for any injury and/or damage to persons or property arising from any methods, products, instructions, ideas or otherwise contained in this publication.

This publication is designed to provide accurate and authoritative information with regard to the subject matter covered herein. It is sold with the clear understanding that the Publisher is not engaged in rendering legal or any other professional services. If legal or any other expert assistance is required, the services of a competent person should be sought. FROM A DECLARATION OF PARTICIPANTS JOINTLY ADOPTED BY A COMMITTEE OF THE AMERICAN BAR ASSOCIATION AND A COMMITTEE OF PUBLISHERS.

LIBRARY OF CONGRESS CATALOGING-IN-PUBLICATION DATA

Syngas generation from hydrocarbons and oxygenates with structured catalysts / authors, Vladislav Sadykov ... [et al.].
       p. cm.
 Includes bibliographical references and index.
 ISBN 978-1-60876-323-8 (softcover)
 1. Synthesis gas. 2. Oil gasification. 3. Hydrocarbons--Oxidation. 4. Catalysts. I. Sadykov, Vladislav.
 TP360.S93637 2009
 665.7'7--dc22

                    2009036572

Published by Nova Science Publishers, Inc. †*New York*

CONTENTS

PREFACE

Syngas generation by oxidative or steam reforming of hydrocarbons and oxygenates now attracts a lot of attention of researchers in the fields of heterogeneous catalysis and chemical engineering due to its tremendous importance for energy generation and synfuels production. This book reviews results of the long-term research of the international team of scientists aimed at development of efficient processes of syngas generation in structured catalytic reactors. Multiscale integrated optimization approach is applied throughout of this work including design of nanocomposite active components stable to coking and sintering; developing heat-conducting monolithic substrates comprised of refractory alloys and cermets (honeycomb and microchannel structures, gauzes etc) and procedures of their loading with active components; design and manufacturing of several types of pilot-scale reactors (with the radial or the axial flow direction) equipped with unique liquid fuel evaporation and mixing units and internal heat exchangers. Extended tests of these reactors fed by fuels from C1 to gasoline, mineral and sunflower oil have been carried out with a broad variation of experimental parameters including stability tests up to 1000 h. Performance analysis has been made with a due regard for equilibrium restrictions on the operational parameters. Transient behavior of the monolith reactor during start-up (ignition) of the methane partial oxidation to synthesis gas was studied and analyzed via mathematical modeling based upon detailed elementary step mechanism. This provides required bases for theoretical optimization of the catalyst bed configuration and process parameters.

INTRODUCTION

Most of the current energy supply system, in which fossil fuels are used, causes many environmental problems: air pollution, acid and greenhouse gases emissions. However, the modern life is strongly related to mass consumption of fuels for electricity generation and transportation sector. These sobering trends justify an urgency with respect to the need for more efficient, as well as alternative, energy conversion technologies and devices. The features, which are required from alternative energy conversion technologies, include high energy conversion efficiency, environmental friendliness, compatibility with both conventional fuels and renewable energy sources and sustainability [1-3].

Hydrogen's unchallenged potential to reduce greenhouse gases and atmospheric pollutants makes it a major candidate to meet the world energy demand, especially for small-scale applications. This simplest element and the most plentiful gas in the universe does not exist alone in nature. It always combines with other elements such as oxygen and carbon, and is mostly present in water, biomass, and fossil hydrocarbons. Conversion of fuels into hydrogen-rich product streams is commonly referred to as fuel processing. The fossil fuels, petroleum, coal and natural gas are high energy density materials and, thus, amenable to centralized, large-scale processing plants. Industry generates ca 50 millions metric tons of hydrogen globally each year from fossil fuels [2-6]. It is expected that significant development of a hydrogen transportation infrastructure will not occur within the next decade. The lack of safe, efficient and cost effective hydrogen storage system is the major obstacle for hydrogen application as an energy carrier [6]. The most promising approach for distributed small/middle-scale applications is the small -scale processing of fuels near the point of usage. Natural gas and liquid fuels such as alcohols, gasoline, jet fuel, kerosene and diesel, are favored as the most suitable hydrogen carriers because they are easily transportable and have existing infrastructures for distribution [7-9]. Nowadays, the modern biomass-based transportation fuels (liquid or gaseous) such as bioethanol, fatty acid (m)ethylester, biomethanol, biodiesel as well as diesel produced from biomass through syngas-Fischer-Tropsch route are generally considered as offering many advantages including sustainability, reduction of greenhouse gas emissions, regional development, social structure and agriculture, security of supply [10].

One of the ways to generate hydrogen-rich gases is catalytic reforming of carbonaceous energy carriers to produce syngas. In its simplest form, syngas is composed of hydrogen and carbon monoxide. Syngas is considered as an alternative to conventional fuels in all its applications. It can be used like natural gas, as a source of hydrogen for fuel cells, or reformed into other hydrocarbon fuels. In principle, syngas can be produced by reformation from any fossil fuels, including natural gas, naphtha, residual oil, petroleum coke, coal, and biofuels. The process is carried out in the presence of a catalyst which controls the product composition. The catalytic reactions take place at the solid-gas or solid-liquid interfaces. The catalyst plays a major role in enhancing the hydrogen yield and reducing the formation of undesired compounds. New catalytic processes hold the greatest promise for increasing the efficiency and environmental benefits of energy related processes [11, 12].

Currently, one of the limitations of hydrogen generation is that it is based on fossil fuels, leading to a net production of greenhouse gases. Thus, the use of biomass as an alternative hydrogen source, with CO_2-neutral impact, may become important in the future. An obstacle to converting biomass into syngas, at an economical scale, has been the decentralized nature of biomass production. Biofuels resulted from conversion of biomass of a different origin would have been produced in a number of small plants. Thermochemical treatment of biomass can produce bio-gases, liquids and solids, yields being determined by such process parameters as the residence time, heating rate, and temperature. Transporting the dense biofuels is cheaper than transporting biomass. Thus, liquid oils, called bio-oils represent a type of easily transported high-energy density chemicals.. Commercially, bio-oils are used as boiler fuel for stationary power and heat production, and for chemical production. Since the production of bio-oils is considered now to be a mature technology, their catalytic treatment (steam/autothermal reforming) on-site to produce hydrogen can be the promising technology for a clean and renewable hydrogen generation [12, 13]. The bio-oils could be treated directly as a whole or using specific fractions. Nevertheless, direct feeding of raw bio-oil into the reformer reactor is not easy, since it is only partially soluble in water and highly unstable upon heating, polymerizing at temperatures as low as $80^{\circ}C$. Fast pyrolysis liquids are complex mixtures of oxygenated compounds, which include alcohols, acetone, and aldehydes, emulsified with water. Such bio-fuels may contain acids, aldehydes, ketones, alcohols, glycols, esters, ethers, phenols and derivatives, as well as carbohydrates, and a large proportion (20–30 wt.%) of lignin-derived oligomers [14, 15]. Thus, different molecules in the bio-oils have a significant difference in reactivity and coke formation rates. It is highly desirable that the oil fractions that lead to thermal coking (such as aldehydes, oxyphenols, and furfurals) be removed from the bio-oil prior to catalytic upgrading. Recently, a lot of research have been carried out in the field of steam reforming of some oxygenated molecules, such as methanol, acetic acid, ethanol, acetone, phenol, cresol etc used as model compounds of bio-oils [16-20].

From the thermodynamic point of view, syngas production from hydrocarbons or oxygenates may be categorized into two basically different types of processes. One is endothermic steam reforming, the other is exothermic partial oxidation, where the feed reacts directly with air, or enriched air with carefully balanced oxygen to fuel ratios. The gross oxidative reaction of syngas formation from a general fuel (hydrocarbon or oxygenate) can be described by the following expression:

$$C_nH_mO_p + x\ O_2 + (n-2x-p)\ H_2O = n\ CO + (n-2x-p+(\tfrac{1}{2}\ m))\ H_2$$

Criteria for operating modes can be estimated with thermodynamic calculations. Nevertheless, the risk of residues and soot formation increases with the complexity of the reformed fuel. For a practical consideration of hydrogen-rich gas production from the particular fuel, the operational characteristics should be chosen on the bases of analysis of many factors including a range of throughputs, design of reactor and configuration of catalytic bed, as well as catalytic activity. The operational parameters have to be optimized to perform the process safely and escape all the possible problems arisen.

The various levels of a catalytic process development are not independent. They must be considered in an integrated approach and in close relationship with the reactor design. The question of the optimal catalyst design and its relationship with the reactor design and feed preparation/supply system is more critical in a small-scale application. There has been a significant interest since the first publications [21-28] in the new catalytic processes, which operate at high temperatures and under short contact times over structured catalysts. Structured catalysts are ceramic and metallic configurations which constitute both the catalyst support and the reactor. The most important structured reactors are based on gauzes, foams and monoliths. The catalyst bed may also be arranged as either arrays of particles or superimposed sheets, to allow cross-flow. They are compact reactors with excellent performance in activity and selectivity. In general, the most satisfactory structured reactors are equipped with monolithic catalysts. Traditionally, monolithic catalysts are continuous unitary structures which contain very small, mostly parallel passages. The catalytically active material is supported as a thin layer (washcoat) over the ceramic or metallic substrate.

In gas-phase applications, the structured reactors are often preferred due to their favourable properties with respect to selectivity, pressure drop and robustness. Their millisecond characteristics have potential in syngas production and selective conversions into valuable products. There is a growing need to extend our catalyst and reaction engineering knowledge base to the structured reactor category, to have better understanding of flow contacting, multispecies transports and detailed reaction kinetics and their interactions in the reactors, to have improved understanding of the catalyst structure impact and materials (washcoat) influence on the process performance [29-33]. Catalytic monoliths with a high thermal stability and thermal shock resistance are also required. Preheat temperature, flow rate, oxygen-to-carbon and water-to carbon ratios in the feed affect the temperature distribution in the structured catalyst. Hence, composition of the active components, configuration of the catalyst bed and the process performance as a whole require optimization for the particular case of the fuel being reformed.

This review considers the important aspects of syngas generation development process from a variety of fuels on structured catalysts with a due regard for results obtained within represented by authors international broad-scale collaboration as well as related world-wide research published in available sources.

Chapter 1

REFORMING CHEMISTRY

The operation mode for the reformer can be very different, with wide implications on the composition of the reformed effluent and the energy demand, necessary to generate the hydrogen-rich feed. As for the reforming chemistry, there are three possible operational modes of the small-scale fuel reforming: catalytic steam reforming (SR), direct partial oxidation (PO) and so called indirect partial oxidation, i.e. combination of partial oxidation and steam reforming in stand-alone catalytic system, viewed as autothermal reforming (ATR) [27, 34-38].

1.1. STEAM REFORMING

Steam reforming involves the reaction of steam with the fuel in the presence of a catalyst, as illustrated in Eq. (1) for methane and Eq. (2) for generic hydrocarbon fuel. This process is the most highly developed and cost effective approach for generating hydrogen and is also the most efficient, giving conversion rates of 70 to 80 per cent.

Reaction (1) $CH_4 + H_2O \leftrightarrows CO + 3H_2$
$\Delta H^{\circ}_{298} = +206.2$ kJ mol^{-1}

Reaction (2) $C_nH_m + nH_2O \rightarrow nCO + (n+m/_2)H_2$

For a generic oxygenated molecule, the steam reforming reaction proceeds according to the following equation:

Reaction (3) $C_nH_mO_k + (n-k)H_2O \rightarrow nCO + (n+m/2-k)H_2$

The steam reforming reactions (Eqs. 1- 3) are followed by the exothermic water gas shift reaction (Eq. 4) and methanation reaction (Eq. 5):

Reaction (4) $CO + H_2O \leftrightarrows CO_2 + H_2$
$\Delta H^{o}_{298} = -41.2$ kJ mol^{-1}

Reaction (5) $CO + 3H_2 \rightarrow CH_4 + H_2O$
$\Delta H^{o}_{298} = -206.2$ kJ mol^{-1}

When the steam-to-carbon ratios are close to the stoichiometric value (Eqs. 1-3), coke produced by thermal cracking of hydrocarbons (Eq. 6) or by the Bouduard reaction (Eq. 7) may lead to the catalyst deactivation.

Reaction (6) $CH_4 \leftrightarrows C + 2H_2$
$\Delta H^{o}_{298} = +75$ kJ/mol

Reaction (7) $2CO \leftrightarrows C + CO_2$
$\Delta H^{o}_{298} = -172$ kJ/mol

To minimize coke formation, excess steam is used to ensure that any carbon formed is gasified

Reaction (8) $C + H_2O \leftrightarrows CO + H_2$.

The steam-to-carbon ratio and operating temperatures depend on the type of fuel. Heavier feeds require a higher steam-to-carbon ratio and a higher operating temperature. For methane, a steam-to-carbon ratio of ~ 2.5 is sufficient to avoid coking. For higher hydrocarbons, a steam-to-carbon ratio of 6-10 is common [5, 39]. While bio-oils are more reactive than petroleum fuels, high temperature is needed in the reactor to gasify the formed coke deposits. High ratios of steam to carbon (greater than 7) are necessary to avoid catalyst deactivation by coking [14].

The thermal decomposition (Eq. 9) has also to be considered for most of the hydrocarbon and oxygenated fuels [40-42]. Steam reforming of bio-fuels is more complicated since some bio-fuels components are thermally unstable and decompose upon heating.

In this case, the carbon is essentially a soot-like deposit.

Reaction (9) $C_nH_mO_k \rightarrow C_xH_yO_z + carbon + (H_2, CO, CO_2, CH_4, C_xH_{x+2})$.

The formed C_xH_y fragments become also favorable for coking via cyclization processes. Deactivation of the catalysts due to coking is one of the major problems, and bio-oils reforming encounters more severe deactivation problems than do petroleum-derived feedstocks. In fact, steam reforming of bio-oils in fixed bed reactors requires a catalyst regeneration step after 3-4 h of time-on-stream [14].

The main disadvantage of this method for hydrogen or syngas production is that it is endothermic and, hence, requires external heating and overheated steam. This reduces the overall efficiency of the system from the theoretical 100 per cent and also means that there is a delay before the system is ready to operate. As a result, this method is not particularly suitable for portable applications as consumers expect devices to start immediately [37].

1.2. PARTIAL OXIDATION

Partial oxidation (also called direct partial oxidation) involves the reaction of the hydrocarbon with oxygen to liberate hydrogen as presented by Eqs. 10, 12:

Reaction (10) $CH_4 + \frac{1}{2} O_2 \leftrightarrows CO + 2H_2$
$\Delta H^o{}_{298} = - 35.7$ kJ mol^{-1}

Reaction (11) $C_nH_m + n/2\, O_2 \rightarrow nCO + m/2\, H_2$

Partial oxidation can be used for converting methane and higher hydrocarbons into syngas but is rarely used for oxygenates [43-50]. The catalytic partial oxidation is the mild exothermic reaction. The oxygen-to-fuel ratio determines the heat of reaction and the hydrogen yield. The partial oxidation processes are attractive alternatives because they avoid the need for large amounts of expensive superheated steam. Direct partial oxidation catalytic reaction in monolithic reactors proceeds at millisecond contact times, which are at least 2 orders of magnitude shorter than those for traditional steam reforming process (~ 1 s), and a high conversion is achieved. However, less hydrogen is produced for the same amount of fuel than in the case of steam reforming. One more advantage of partial oxidation is that formation of olefins via endothermic catalytic dehydrogenation and cracking is excluded due to a high conversion at the short residence time. A low coking was attributed to the fact that carbon-producing reactions (olefin cracking, CO disproportionation, and reverse-steam reforming) do not approach equilibrium and that the presence of CO_2 and H_2O favor reforming of the carbon formed [43].

1.3. AUTOTHERMAL REFORMING

This process (so called indirect partial oxidation) combines the endothermic steam reforming process with the exothermic partial oxidation reaction (Eqs. 12, 13) in stand-alone catalytic system, therefore balancing the heat flow into and out of the reactor [7, 9, 34, 37, 51-54]. No external heating source is required.

Reaction (12) $2 CH_4 + \frac{1}{2} O_2 + H_2O \leftrightarrows 2CO + 5H_2$
$\Delta H^{\circ}_{298} = -18.4$ kJ mol^{-1}

The general formula for ATR, using air as the oxygen source is

Reaction (13) $C_nH_mO_p + x(O_2 + 3.76N_2) + (n-2x-p) H_2O = nCO + (n-2x-p+(\frac{1}{2} m)) H_2 + 3.76xN_2$.

The fuel gas contains a mixture of H_2, CO, CO_2 with the relative concentrations being determined by the water-gas shift reaction (Eq. 4), if thermodynamic equilibrium is achieved. A lower operating temperature of the catalytic ATR compared to PO has a several advantages for small-scale applications [46]:

- less complicated reactor design and lower reactor weights, because less thermal integration (i.e. heat exchange between incoming reactants and hot products) is required;
- a wider choice of materials required;
- a lower fuel consumption during start-up, because for a given reactor mass the energy required to heat the reformer to its operating temperature is proportional to its operation temperature.

The amount and concentration of hydrogen generated from a given amount of hydrocarbon fuel, and the quality of the raw reformate (CO, CO_2, CH_4 and possibly other species, H_2O and nitrogen contents) are determined by reforming conditions.

1.4. COKE DEPOSITION

Coke deposition takes place where polymerization, thermal decomposition of fuel and other reactions occur, leading to blockage of catalyst pores and, in extreme cases, to complete failure of the reactor. The reaction conditions must be maintained such that no graphitic or amorphous carbon is formed in the reformation process. Depending on the catalyst used, the reforming reaction may require temperatures of $\sim$1000-1300 K, or even higher. At these temperatures, most of the carbon in the fuel is converted to CO and CO_2, with possible formation of relatively small amounts of CH_4 and other hydrocarbons. Carbon deposition in the reactor may also take place by the direct decomposition of methane and by the Boudouard reaction (reactions 6 and 7). Carbon formation is a problem in reforming of hydrocarbon fuels, particularly for hydrocarbon fuels with two or more carbon atoms in the main chain. More aromatic fuels have a higher tendency of carbon formation. Heavier hydrocarbons in the jet fuels and diesel fuels can form carbon deposits at relatively low temperatures such as 450°C due to fuel pyrolysis [10, 55, 56]. A catalyst applied may initiate carbon formation as well. For example, carbon formation catalyzed by Ni usually takes place in the form of whiskers with a Ni particle at the top of a fibre [57].

Steam reforming of oxygenated hydrocarbons is thermodynamically favored at lower temperatures than that of hydrocarbons [58], so the steam reforming of oxygenates can take place at lower temperatures because it is less endothermic [59]. Two main reactions, steam reforming and pyrolysis, occur in a high temperature steam/bio-fuel mixture. However, many reactions occur simultaneously in the reformer including many side reactions. Side reactions provide a higher thermodynamic stability of the overall process operation by coupling exothermic and endothermic reactions. It was found that to prevent carbon deposition, the oxygen/carbon ratio (O/C) in the reaction system have to satisfy the following requirements: in pyrolysis O/C ≤ 1, gasification O/C < 2 and combustion O/C > 2 [60]. An equilibrium phase diagram (Figure 1) taken from Slinn et al [60, 61] represents graphically gasification and reforming systems. The phase diagram shows that, above the carbon deposition boundary (dashed line), solid carbon particles exist in equilibrium with the gaseous components. Below this carbon deposition boundary (shaded section) carbon is present as CO, CO_2 and CH_4. To avoid additional carbon formation either oxygen or H_2 must be added to shift the point below the defined carbon deposition boundary. Further addition of oxygen will shift the equilibrium position over the line of complete carbon combustion where free oxygen is present [60, 61]. Fossil fuels have a lower oxygen content within their molecular structure, and they would be placed above the carbon boundary in equilibrium with solid carbon. Such oxygenate as glycerol contains a higher oxygen content and has an oxygen/carbon ratio of 1, so it is already at the carbon boundary and does not need any extra oxygen or hydrogen. However, this phase diagram assumes thermodynamic equilibrium, which might not always be the case. Therefore, there may be more carbon than predicted.

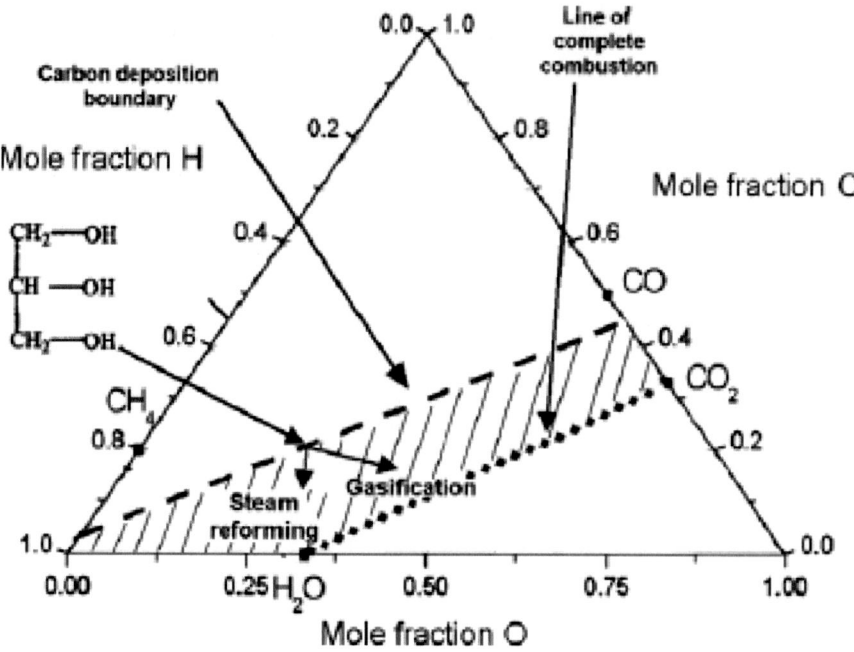

Figure 1. Carbon–hydrogen–oxygen equilibrium phase diagram. Adapted with permission from Ref. [60]. Copyright 2008 Elsevier B.D.

Actually, there are two limits to be considered for the coke deposition in the hydrocarbons reforming: thermodynamic and kinetic. By thermodynamic criteria, coke is formed if the gas composition shows affinity for carbon. From kinetic point of view, carbon formation in the reactor occurs in conditions where hydrocarbon can decompose yielding carbon, even if thermodynamics predicts no carbon formation after the equilibrium has been reached. Carbon formation is then a question of kinetics, local process conditions and reactor (reformer) design [57, 61-63].

Chapter 2

CATALYSTS FOR REFORMING HYDROCARBON FUELS AND OXYGENATES

The small- scale processing of hydrocarbon fuels and oxygenates to hydrogen-rich gas requires that the reforming catalyst exhibit a higher activity and better thermal and mechanical stability than the reforming catalysts currently used for the production of H_2 in large-scale industrial processes.

2.1. CATALYSTS FOR SYNGAS GENERATION FROM HYDROCARBON FUELS

Catalyst are typically group VIII metals, such as rhodium, platinum, palladium, ruthenium, cobalt, nickel, and iridium [64-71], which are either supported on oxide substrates [64, 65], or used unsupported, as metal wires and gauzes [66] for partial oxidation of hydrocarbons to hydrogen-rich gas. Au et al. [67] carried out a theoretical study on the comparison of single component catalysts such as Rh, Ru, Ir, Os, Pd, and Pt, and concluded that the most efficient catalyst for methane dissociation is Rh, which results in a high CH_4 conversion. However, at stoichiometric fuel-to-oxygen ratio, a high and stable performance was demonstrated only for catalysts with a high (5-10 wt.%) loading of expensive Rh on corundum foam substrates, while less expensive supported Pt/alumina catalysts were less active and selective [22-26, 45-48] deactivating with time-on – stream due to coking [68-70]. Moreover, in the short contact time monolithic reactors all oxygen in the feed is consumed in a narrow (1-2 mm) inlet part of the catalytic layer, where high (up to 1200°C) temperatures are developed due to oxidation reactions [22]. As the result, evaporation and loss of precious metals followed by continuous moving of the high-temperature zone along the monolith length may occur up to complete deactivation of the catalyst.

From the work of M. Prettre et al. [71] which can be considered as a historic benchmark for research on the catalytic partial oxidation of methane, the family of the inexpensive catalysts on the base of Ni supported on refractory Mg-Al spinel is widely used in industry [57] due to their high stability and activity in synthesis gas production. The study by Vermeiren et al [72] on the catalytic activity of Ni supported on Al_2O_3, SiO_2-Al_2O_3, SiO_2- ZrO_2, and zeolite Y shows that complete oxidation followed by steam or CO_2 (or both) reforming reactions could occur on all these types of catalysts. The Ni- supported group of catalysts on the bases of rare-earth (La, Ce, Pr, Nd, Sm, Eu, Gd, Dy, Er, Yb) oxides show a similar activity and selectivity for the oxidation reforming reaction [73]. The tendency to coke deposition is generally considered to be the main drawback in the application of supported Ni catalysts. Among the catalyst group mentioned, NiO-La_2O_3 catalyst was the best one. The addition of rare earth metal oxide or alkaline metal oxide to alumina or the use of rare-earth oxides as support can somewhat restrict carbon deposition. The promotion by the rare earth oxide addition to the catalyst support is probably due to its capability to act as oxygen or oxygen - containing compounds storage, which can help in oxidizing the deposited coke or its precursors. Moreover, decoration of the surface of Ni particles by hydroxocarbonate species could prevent carbon nucleation due to ensemble dilution effect. It is also believed that the presence of rare earth oxide such as CeO_2 can stabilize the support and prevent it from sintering during the high-temperature reaction. Cerium is supposed to weaken a strong bond of Ni with aluminum oxide. This prevents Ni diffusion into the alumina lattice [74, 75].

Perovskites, the composite oxides of alkali-earth (rare-earth) and transition metals with a general formula ABO_3, are the vast class of catalytic systems for oxidation reforming of hydrocarbons. Cation of transition metal B is in the octahedral oxygen surrounding. A big cation A is surrounded by the 12 oxygen ions. A dense packing is provided by AO_3 layers comprised of oxygen anions and A cations. Lattice defects are formed if A or B cations are replaced by cations with different charge. Activity of perovskites in aq lot of catalytic reactions has been intensively investigated since 1970. Catalytic activity seems to correlate with the density of lattice defects [76]. In the case of nickelates as catalysts of steam reforming or partial oxidation reactions, metallic Ni formed under reducing reaction feed effect is capable to provide the homolytic rupture of C-H bond followed by coke formation from activated CH_x fragments if their subsequent transformation into syngas is a slow step. In this case, triple-charged La and Cr cations in the lattice of complex perovskites improve the oxygen anions mobility in the lattice and stabilize Ni in the state of 1+ or 2+ [77, 78].

The most promising type of the active component is a nanocomposite containing precious metals combined with mixed oxides possessing a good oxygen mobility and reactivity. Thus, catalysts developed at Argonne National Laboratory and produced by Süd-Chemie Inc. contain a transition metal supported on an oxide-ion-conducting substrate, such as ceria, zirconia or lanthanum gallate doped with a small amount of non-reducible element, such as gadolinium, samarium etc [34, 37]. Various transition metals supported on doped ceria exhibited excellent isooctane reforming activity between 500 and 800°C with a high fuel conversion and H_2 selectivity. Among the metals investigated (Ni, Co, Ru. Pd, Fe, Cu, Ag), all metals except for Ag, exhibit conversion of > 95 % at T > 600°C, and all metals exhibit 100 % conversion at 700°C. At temperatures <600°C, conversion decreases more rapidly for the first-row transition metals (in particular, Ni and Co), than for the second row (Ru) and the third-row (Pt, Pd) transition metals. The second- and third-row transition metals exhibit a higher selectivity to H_2 (>60%) than the first-raw transition metals at T > 650°C. At temperatures < 600°C, the H_2 selectivity decreases to <50% for all metals except Ni and Ru [79, 80].

Recent research also demonstrated that active components comprised of precious metals (platinum, rhodium) supported on complex oxides with the fluorite-like structure (doped ceria or ceria-zirconia) or complex perovskites (nickelates etc) are efficient in selective oxidation of methane into syngas [81-87]. The main point in the catalytic systems developed for oxidative reforming of hydrocarbon fuels to syngas is that a part of precious metal could be replaced by less expensive metal such as Ni, thus decreasing a cost of the catalyst. Nickel having a rather high ability to activate methane is attractive as a component for the hot (temperature 1100-1200 °C) inlet part of the monolithic catalyst, since in the presence of oxygen NiO or $LaNiO_3$ possesses a much lower volatility as compared with PtO_2. Ni-containing complex oxides promoted by small amounts of precious metals show excellent performance in the autothermal reforming of hydrocarbons at short contact times with low ignition temperatures [68]. They were demonstrated to be reasonably efficient in methane partial oxidation [84-88] as well as steam [68, 89-93] and dry [94-100] reforming. Under contact with the reducing reaction feed at high operation temperatures, Ni-containing complex oxides are transformed into small clusters of metallic nickel strongly interacting with the mixed oxide support which are responsible for catalytic performance, while big Ni particles weakly interacting with support are rapidly deactivated due to coking [100]. Certainly, performance of these catalytic systems strongly depends both upon the content of Ni, nature of complex oxide support/matrix, its structural features as well as upon the nature and content of promoting platinum group metals, which requires systematic studies of all these factors for optimization of the composition. The strongest positive effect is observed when Pt-Ni alloys or surface La-Ni-Pt-O perovskites are formed.

While two-component ceria-based systems doped with Sm, Gd, Pr, Y, La etc are thought to deactivate supported precious metals in reducing conditions due to alloying, decoration or electronic effects [101, 102], ceria-zirconia solid solutions appear to be free of those shortcomings [102]. On the other hand, ceria—zirconia solid solutions suffer from decomposition at high temperatures in hydrothermal conditions into phases enriched, respectively, by Ce and Zr [103]. High-performance active components of catalysts for selective oxidation of hydrocarbons (gaseous and liquid) into syngas at short contact times comprised of nanocrystalline ceria-zirconia solid solutions doped by La, Pr etc prepared via polymerised polyester precursors (Pechini) method [104] and promoted with a small amount of noble metals (~ 1 wt.%) were developed at the Boreskov Institute of Catalysis [81, 82].

In combination of platinum with ceria-(ceria-zirconia) containing fluorite-like complex oxides, Pt is mainly involved in activation of hydrocarbon molecule by the homolytic rupture of C-H bond thus generating hydrogen atoms and CH_x fragments adsorbed on the metal particles [68, 70]. Pt is considered to be the most efficient among Pt group metals in activation of methane molecule [105].

Ceria-zirconia is one of the main components of current generation of three-way automotive catalysts [106] and transformation of hydrocarbons or oxygenates into syngas by partial oxidation or autothermal reforming [81, 82, 107], solid oxide fuel cells anodes [108] etc. At high (400–800°C) temperatures, the redox cycle $Ce^{3+} \Rightarrow Ce^{4+} + e^-$ facilitates the oxygen storage and release from the bulk fluorite lattice. When combined with noble or non- noble metal particles, this makes them ideal candidates for catalytic oxidation applications such as partial oxidation of hydrocarbons into syngas [81, 82, 87-89, 107]. Incorporation of zirconium into the ceria lattice creates a high concentration of defects improving, thus, the oxygen mobility and oxygen storage capacity [109]. The main features which contribute to the success of ceria-zirconia are as follows:

- A higher sintering resistance as compared with CeO_2;
- A higher reduction efficiency of the redox couple Ce^{4+} - Ce^{3+};
- Good oxygen storage/release capacity.

As was recently shown by combination of transient techniques (TAP and kinetic relaxation studies at atmospheric pressure) [110, 111], a high efficiency of Pt –supported doped ceria-zirconia catalysts in partial oxidation of methane at short contact time is determined by realization of direct route of methane pyrolysis- selective oxidation into syngas [112]. Thus, for both black Pt and Pt supported on alumina, small pulses of methane supplied on the oxidized surface produce only combustion products –CO_2 and water (Figure 2).

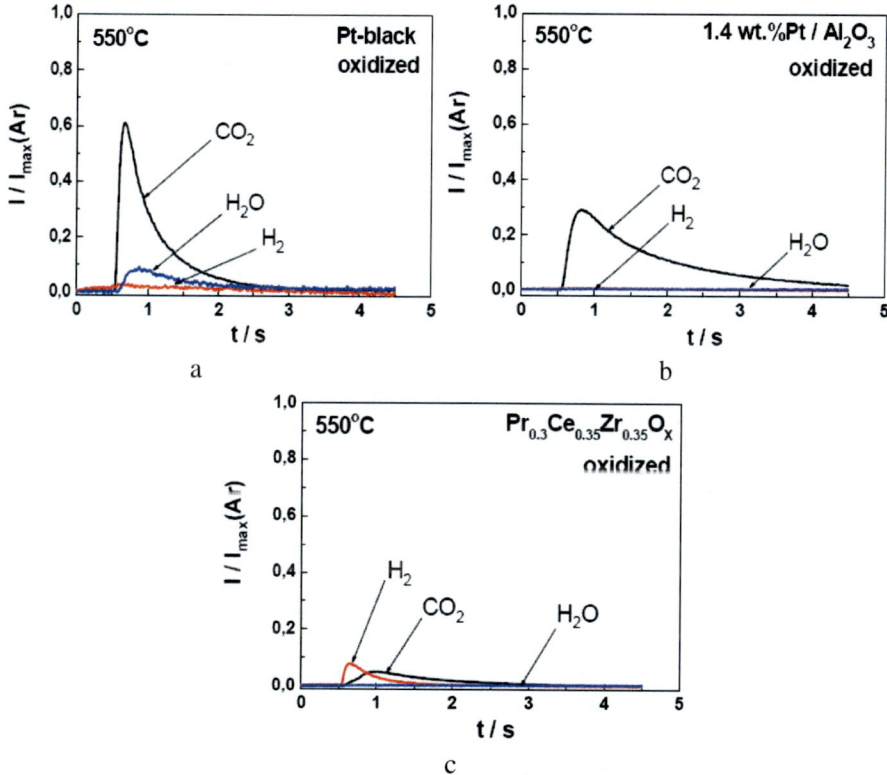

Figure 2. Carbon oxide, carbon dioxide, hydrogen and water pulse responses recorded in 1 part oxygen - 2 parts methane pump-probe TAP experiment at 550°C after dynamic oxygen saturation over Pt -black (a), 1.4 wt.-% Pt/Al_2O_3 (b) and Pr-Ce-Zr-O oxide (c) catalysts [111].

For Pr-doped ceria-ziconia oxide, even without supported Pt, CH_4 pulsing produces some amount of syngas, though in a small amount. For Pt-promoted Pr-Ce-Zr oxide supported on the wall of separate corundum triangular channel [110, 111], vacuum TAP experiments revealed formation of syngas in CH_4 pulses supplied on the partially oxidized surface (Figure 3) and even with a short time lag between O_2 and CH_4 pulses (Figure 4). This indicates that H_2 is indeed a primary product of methane activation on this nanocomposite active component.

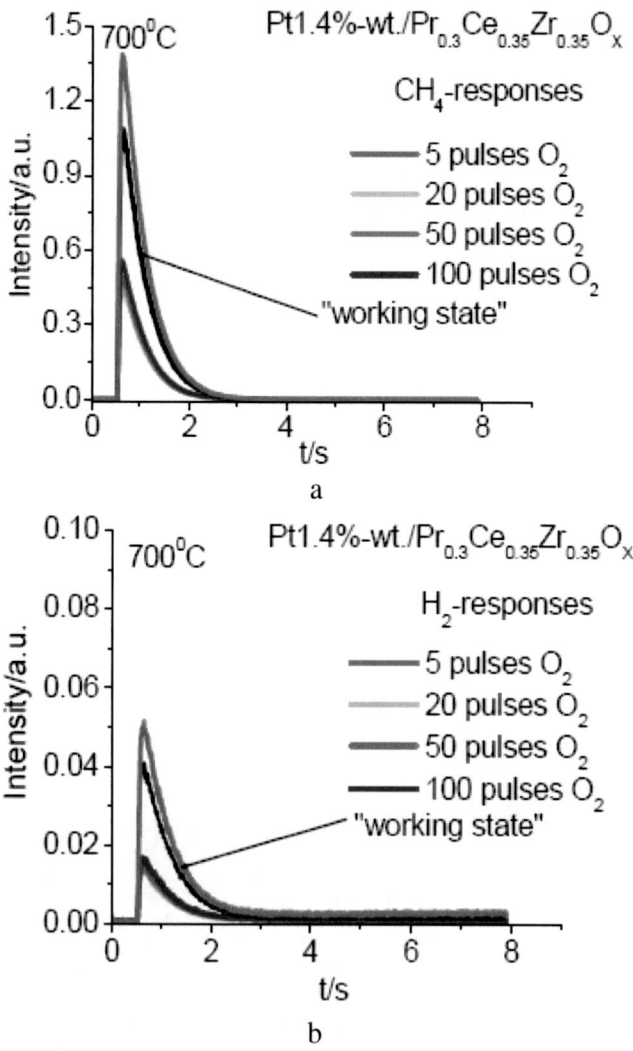

Figure 3. Methane (a) and hydrogen (b) pulse responses over a 1.4 wt% $Pt/Pr_{0.3}Ce_{0.35}Zr_{0.35}O_x/Al_2O_3$ recorded in 1 part oxygen- 2 parts methane pump-probe TAP experiment at 700°C after 5-100 oxygen pre-pulses and in "working" state obtained after a long train of O_2-CH_4 pulses [111].

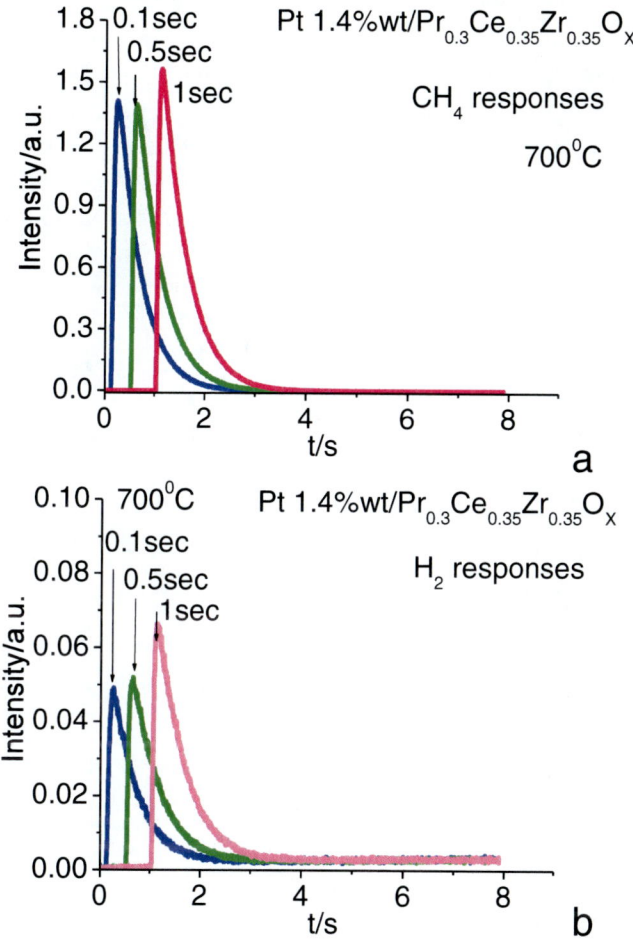

Figure 4. Methane (a) and hydrogen (b) pulse responses over a 1.4wt.%Pt/$Pr_{0.3}Ce_{0.35}$ $Zr_{0.35}O_x$/Al_2O_3 recorded in 1 part oxygen- 2 parts methane pump-probe TAP experiment at 700°C after 5 oxygen pre-pulses with different time lag (0.1-1 s) between pulses of O_2 and CH_4 [111].

Similarly, for concentrated feeds, transient kinetic experiments with separate corundum thin-wall channels with supported doped ceria-zirconia complex oxides promoted by Pt revealed that after switching He stream to the CH_4 +O_2 reaction feed, for oxygen –pretreated Pr-doped catalyst, H_2 and CO appear simultaneously with CH_4 at the reactor exit [110, 111] (Figure 5). This proves that direct pyrolysis-selective oxidation route of syngas generation operates for the real concentrated feed as well. For Gd-doped sample some delay in H_2 appearance was explained by accumulation of surface intermediates (formates etc), which are decomposed yielding H_2 and CO [110]. For both samples concentration of syngas goes through the maximum at ~ 100 s (Figure 6).

In agreement with TAP studies results, this suggests that partially oxidized surface is the most efficient in syngas generation. This feature can be explained both by a lower efficiency of oxidized Pt clusters in CO and H_2 combustion as well as by partial blocking of supported Pt and/or neighboring centers of complex oxide support by carbonaceous species.

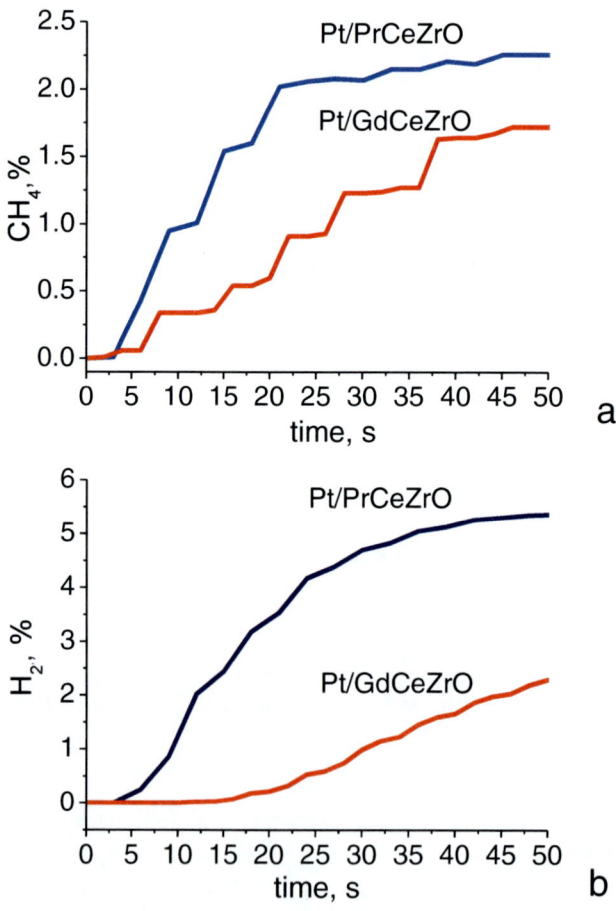

Figure 5. Initial CH_4 (a) and H_2 (b) transients after switching He stream for the mixture of 7% CH_4 +3.5% O_2 in He. Oxidized samples of Pt/PrCeZrO and Pt/GdCeZrO catalysts; contact time 15 ms, 650°C [111].

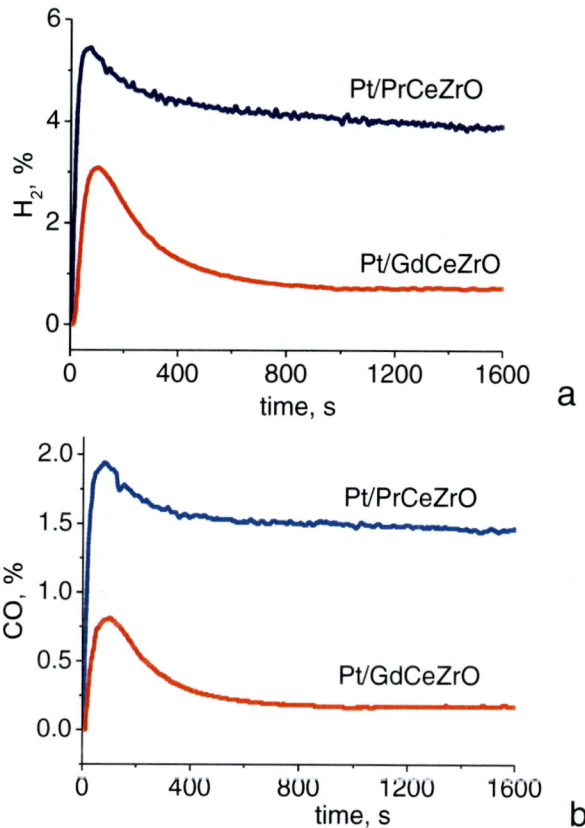

Figure 6. Slow transients of H_2 (a) and CO (b) concentrations at the reactor exit after switching He stream for the mixture of 7% CH_4 +3.5% O_2 in He. Oxidized samples Pt/PrCeZrO and Pt/GdCeZrO catalysts; contact time 15 ms, 650°C [111].

Such deactivation is counteracted by the oxygen species migration from support (where they are produced by O_2 dissociation on oxygen vacancies) to the Pt clusters perimeter where they are involved in gasification of CH_x species produced by CH_4 dissociation. A higher lattice oxygen mobility in complex fluorite-like oxide support found for Pr-doped sample [110, 111] indeed correlates with both a faster relaxation and a smaller activity decline for this catalyst (Figure 6). Further details of kinetics, mechanistic and physico-chemical characterization of Pt/ceria (ceria-zirconia)-based materials in methane partial oxidation are reported elsewhere [113-123]. Compositions and methods of these catalysts preparation are covered by the patent of Russia [124].

2.2. Catalysts for Transformation of Oxygenates into Syngas

Design of efficient, inexpensive and robust catalysts for syngas production from bio-fuels is even more challenging than that for CH_4 selective oxidation. Severe coking and deactivation is a major problem for both biomass and bio-oil gasification and steam reforming even in the feeds with excess of steam [125-133]. Transformation of some intermediates (e.g., acetone etc), formed during reaction on the support leads to oligomers and coke formation. These species block the catalytic active sites resulting in deactivation. Such polymerization cannot be avoided [16, 128] on the most common catalysts reported for gasification of biomass such as dolomite, alkali metal and nickel-based catalysts [73], traditional Ni-based steam reforming catalysts [129], Ni–yttria doped zirconia (YSZ) anode cermets [130] or even precious metals (Pt, Pd and Ru) supported on traditional carriers with developed surface area (alumina and zirconia) known to be highly active and selective in syngas generation from methane [132]. Some oxide supports are believed as being responsible for the catalysts coking as well [133].

One of the options to overcome catalyst deactivation is regeneration of the catalyst periodically [16, 41]. In this case, the oligomers/coke accumulated on the catalyst are combusted in air. Thus, a good activity and stability of Ni–K/La_2O_3–Al_2O_3 catalyst in a sequential cracking/reforming reaction using crude bio-oil as feed has been proven by the group of C. Mirodatos in Lyon [41, 42]. The process alternates cracking steps, during which a H_2 + CO rich stream is produced and carbon is stored on the catalyst, with regeneration steps where the carbon is combusted under oxygen. With this process, the cracking steps yielded 45–50% of H_2 in the products stream. However, the discontinuous operation as the reaction is stopped during the regeneration step limits productivity, and catalyst lifetime is too short to do this on a commercial scale.

Supported Ni, Pt, Rh, and Ir catalysts were reported to be effective for the oxygenates reforming reaction [52, 134-140]. To promote syngas formation, ceria, cobalt, and ruthenium are added [52, 141, 142]. Thus, Ir/CeO_2 catalysts show excellent stability for both steam and autothermal reforming of ethanol at relatively low temperatures using approximately stoichiometric feed streams [139, 143]. Rather high stability for the steam reforming, partial oxidation and autothermal reforming of ethanol over Ir/CeO_2 catalysts was demonstrated at 823 and 923 K with no apparent deactivation for 60 h on stream [144]. A conclusion has been made that CeO_2 likely prevented highly dispersed Ir particles from sintering and inhibited coke deposition through strong Ir–CeO_2 interactions. In steam reforming of ethanol, its dehydrogenation into acetaldehyde was found to be a primary reaction [145]. Acetaldehyde is then decomposed to CH_4 and CO. Methane is further reformed to H_2 and CO, and the latter is converted to CO_2 by the water gas shift (WGS) reaction. Ir particles are mainly responsible for breaking C–C bonds, while CeO_2 promotes the activation of water molecules favoring the reforming of CO and CH_4.

Rh/CeO_2 catalyst is known to be suitable for the synthesis gas production by partial oxidation of alkanes [146]. The rhodium–ceria catalyst was applied in [147] to study the catalytic partial oxidation of methyl acetate as a simplified version of biodiesel with a purpose to clarify the role of the methyl ester functional group in conversion to synthesis gas at a short (~0.02 s) contact time. Biodiesel by-product has a longer carbon chain attached to the methyl ester functional group, with a C/O ratio of about 1, and a high boiling point. Cylindrical a-alumina foam monoliths with a diameter of 17 mm, pore density of 80 pores per inch, and porosity of 0.83 served as the support material for the active component. The overall partial oxidation of methyl acetate is in fact an endothermic reaction, as it was reported to be the case of ethanol [52]

$$C_3H_6O_2 + \tfrac{1}{2}\,O_2 \rightarrow 3CO + 3H_2$$

$$\Delta H^\circ_{298} = 178.4 \text{ kJ mol}^{-1}$$

This requires the input of a small amount of excess oxygen in order to have some combustion occurring in the reactor, thus supplying the heat for sustaining the autothermal operation of the reactor.

$$C_3H_6O_2 + 3\tfrac{1}{2}\,O_2 \rightarrow 3CO_2 + 3H_2O$$

$$\Delta H^\circ_{298} = -1496 \text{ kJ mol}^{-1}$$

It was found that the methyl ester functional group may not readily decompose to synthesis gas even over Rh/ceria, which shows a higher conversion and hydrogen selectivity than rhodium alone. The effect of the ceria was not as large for methyl acetate as it is for other fuels. The typical biodiesel was more reactive than methyl acetate.

The performance of Pt/$Ce_{0.5}Zr_{0.5}O_2$ and Rh/$Ce_{0.5}Zr_{0.5}O_2$ catalysts deposited on cordierite monoliths (20 mm diameter, 17 mm length, 1200 cpsi) was studied by Domine et al. [42] in the conversion of biomass-derived crude oil towards H_2

production via both steam reforming and sequential cracking processes. In the case of sequential cracking, the process alternated cracking steps, during which the bio-oil is converted into H_2, CO, CO_2, CH_4 and carbon is stored on the catalyst, with regeneration steps where the deposited coke is burnt under O_2. The catalytic behaviour of both Pt and Rh catalysts was found to be very similar, whereas the higher H_2 productivities were achieved with Pt catalyst than with Rh in steam reforming of bio-oil. However, a significant amount of coke was found to accumulate on both catalysts during the steam reforming reaction. The amount of deposited coke is similar for both catalysts. It was revealed that the amount of added water determines the H_2 yield for both noble metals.

In studies of Seshan, Lefferts et al [148, 149], a bifunctional mechanism has been proposed for Pt-supported catalysts in the reaction of acetic acid (AcOH) steam reforming. AcOH was shown to be activated on Pt via splitting C-C bond producing CH_x species and CO_2, while water is activated on such hydrophilic supports as ZrO_2 etc. Both acetic acid and, especially, acetone at 750°C and very high space velocities are thermally steam reformed and decomposed to a mixture of carbon containing products (CO_x and CH_4) and a low amount of hydrogen. Acetone can form substantial amounts of acetic acid in the presence of steam as the latter seems to promote the alternative reforming (oxidation) of acetone to acetic acid. The amount of coke formed via homogeneous reactions is quite low for two oxygenates. The lowest coke deposition rate was observed over 0.5% Rh supported on CaO $2Al_2O_3$ [54].

Bioethanol, a mixture of water and ethanol produced by fermentation of biomass, is considered as an attractive feedstock for the sustainable production of hydrogen for fueling polymer electrolyte fuel cells. Steam reforming, partial oxidation, and autothermal oxidative steam reforming (or indirect partial oxidation) are effective routes for producing hydrogen from bioethanol. The most effective metal in steam reforming with respect to ethanol conversion and hydrogen selectivity is considered to be Rh [143, 150-152], probably due to its high efficiency in breaking the C–C bond of ethanol

$$CH_3CH_2OH + 3H_2O \rightarrow 2CO_2 + 6H_2.$$

Again, Rh/CeO_2 catalyst was significantly more active and selective at 873–1273 K for the partial oxidation of ethanol as an alternative route for the production of hydrogen [153]

$$CH_3CH_2OH + 1.5O_2 \rightarrow 2CO_2 + 3H_2.$$

The autothermal reforming of ethanol and methanol at contact times in the milliseconds (<10 ms) range is characterized by a high (85%) hydrogen selectivity at conversions exceeding 95% [50, 52]. The autothermal partial oxidation of alcohols was studied by Wanat et al. [49] in short-contact time reactors on α-alumina foam monoliths with supported Rh/ceria active component. All alcohols produce H_2 at 70–90% selectivity, and Rh/ceria was superior to Rh in producing H_2. The selectivity of catalytic partial oxidation of alcohols varies strongly with the type of alcohol and with the catalyst used. For methanol, a high conversion and a high H_2 selectivity were found even at high C/O ratios where operation temperatures fell below 600∘C. For 2-propanol, conversions and H_2/CO selectivity were lower than for other alcohols, and mainly oxygenates and olefins were formed, especially for Rh without ceria. These results show that different alcohols have very different selectivity in catalytic partial oxidation at short contact times even at high temperatures. Rapid adsorption of alcohols as alkoxy species leads to complete dissociation to H_2 and CO. The results suggest that acetone and olefins most likely are produced primarily by homogeneous reactions after consumption of all O_2 in the catalytic layer.

Steam reforming of glycerol and biodiesel by-products has been studied by Slinn et al. in [60, 155] by using 0.5 wt.% Pt/Al_2O_3 catalyst. Steam reforming of glycerol was found to be the dominant mechanism at temperatures above 700°C. Optimum reformer performance was reached at 880°C, flow rate of 0.12 mol (glycerol)/min per kg catalyst and steam/carbon ratio of 2.5. In long running experiments, pure glycerol deposited 0.4% of feed as carbon whereas by-product glycerol deposited 2% of feed.

Water activation is the demanding step in the reaction; enhancement of water sorption/activation is essential to increase the steam reforming activity. Seshan et al. have proposed that the redox oxide could be involved in the activation of steam [16]. Metals are supposed to be involved in the activation of the hydrocarbon and oxygenated molecules, the fragments of which react with the lattice oxygen at the interface of the metal and the redox support [128]. Ni, in contrast to Pt, is able to activate H_2O via formation of NiO [57, 156].

Recently, modification of traditional Ni/yttria-doped zirconia (YSZ) anode cermets with ceria–zirconia fluorite-like solid solutions doped by Pr, La or Gd was shown to suppress their coking in stoichiometric methane– steam feeds, while supporting small (~1 wt.%) amounts of precious metals (Pt, Ru and Pd) allowed to provide a high performance in the intermediate temperature range (500–600°C) at short (milliseconds) contact times [157-159]. Such inexpensive composites could also be attractive as active components of catalysts for steam reforming of biofuels including direct reforming of biofuels on anodes or indirect in-cell reforming in fuel channels of solid oxide fuel cells. To ensure a high performance of these composites in steam reforming of a given type of biofuel component, their composition and preparation procedures are to be properly optimized, first of all, as related to the type and content of doping cations in fluorite-like ceria–zirconia solid solutions controlling mobility and reactivity of the surface and bulk oxygen species [160, 161]. In this respect, application of the combinatorial synthesis procedures using robotic workstations which provide unique possibilities of fast preparation of a large number of samples required for such optimization seems to be very promising [162]. In such a way, using robotic workstation based on the Hamilton Microlab Duo system, a number of composite materials based on NiO/YSZ cermet promoted with doped ceria–zirconia fluorite-like mixed oxides and Ru were synthesized and characterized by XRD, BET, TEM, H_2 TPR by Sadykov et al. [163]. Their catalytic performance was estimated in the reactions of methane, ethanol and acetone steam reforming at short (10–36 ms) contact times. The amount of deposited carbonaceous species was estimated by temperature-programmed oxidation. It was found that complex fluorite-like oxides as promoters minimize coking even in stoichiometric fuel/steam feeds. Promotion by small (<1 wt. %) amount of Ru further decreases coking and facilitates activation of such molecules as CH_4 and acetone, thus ensuring a high performance in the intermediate temperature (500–600°C) range. Factors controlling performance of these composites in steam reforming reactions such as lattice oxygen mobility in complex oxide promoters controlled by their chemical composition, strong interaction between components in composites, state and reactivity of supported Ru were assessed [163]. NiO/YSZ composite co-promoted by Ru and ceria–zirconia solid solutions co-doped with Pr and Sm cations are the most promising systems for oxygenates and methane steam reforming possessing a high performance and stability to coking. Typical results are shown in Figures 7 and 8 for steam reforming of ethanol and acetone. At short contact times CO content much exceeds that of CO_2 being apparently determined by the reaction mechanism [145] and not water gas shift reaction equilibrium.

A number of catalysts based on Al_2O_3 loaded with doped Ce-Zr mixed oxides and different active components (Cu, Cu-Ni, Ru, Pt, etc.) synthesized via standard wet impregnation method with the robotic workstation was tested in steam reforming of ethanol [164]. Ethanol (EtOH) was taken as a model compound of bio-oil and its steam reforming as a model reaction. Activity screening experiments were performed at 600–700°C in the mixture 0.5 % C_2H_5OH + 2.5 % H_2O+97%He. The most effective catalyst composition is $Ru/Ce_{0.4}Zr_{0.4}Sm_{0.2}/Al_2O_3$, which agrees with results of [163].

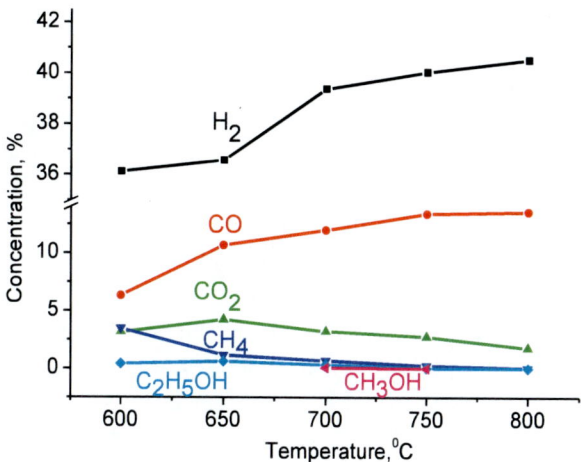

Figure 7. Product distribution for steam reforming of ethanol on nanocomposite 0.9 wt.% Ru/10 wt.% $Pr_{0.15}Sm_{0.15}Ce_{0.35}Zr_{0.35}O_2/(60$ wt.%NiO + 40 wt.%YSZ) sample. Feed composition $EtOH:H_2O:N_2 = 1:4:5.$, contact time 70 ms. Adapted with permission from ref 163. Copyright 2009 Elsevier B.V.

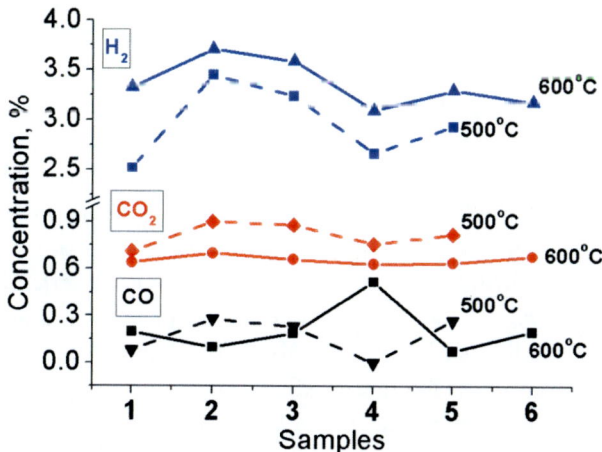

Figure 8. Concentration of products in the steam reforming of acetone for 60 wt.% NiO + 40 wt. % YSZ composites promoted by 10 wt. % of (1) $Pr_{0.15}Sm_{0.15}Ce_{0.35}Zr_{0.35}O_2$; (2) $Ru/Pr_{0.15}Sm_{0.15}Ce_{0.35}Zr_{0.35}O_2$; (3) $Pr_{0.15}La_{0.15}Ce_{0.5}Zr_{0.2}O_2$; (4) $Pr_{0.15}La_{0.15}Ce_{0.35}Zr_{0.35}O_2$; (5)0.9 wt.% $Ru/Pr_{0.15}La_{0.15}Ce_{0.35}Zr_{0.35}O_2$; (6) 1.3 wt.% $Ru/Pr_{0.15}La_{0.15}Ce_{0.35}Zr_{0.35}O_2$. Contact time 20 ms, feed 0.8% $C_3H_6O + 3.5\%$ H_2O in He. Adapted with permission from ref 163. Copyright 2009 Elsevier B.V.

Estimation of catalytic activity at high reagent concentrations (10 % C_2H_5OH + 40 % H_2O + 50 % N_2) at 650–800°C confirmed this fact revealing also that at high temperatures the activity of Cu-Ni catalyst is comparable with that of Ru-containing catalyst. However, the Cu-NiCZ sample was appreciably subjected to coking.

In spite of the fact that methanol steam reforming has been already developed at the industrial level [165], the catalysts with a high thermal stability are still required. Traditional CuO/ZnO or CuO/ZnO/Al$_2$O$_3$ catalysts have a low thermal and long-term stability at temperature near and above 300°C [166-170]. Addition of zirconia to traditional CuO/ZnO/Al$_2$O$_3$ catalyst was shown to considerably improve catalyst stability [168]. In detailed studies of the effect of composition of copper-containing catalysts on their performance in the steam reforming of methanol by Ross et al [169], a series of Cu–zirconia catalysts containing various additives (Y_2O_3, La_2O_3, Al_2O_3 and CeO_2) have been prepared by coprecipitation and their activities and stabilities under operating conditions have been characterized. It has been found that an yttria promoted catalyst containing 30 mol% Cu and 20 mol% of Y_2O_3 is not only very active but is also very stable under reaction conditions. The yttria appears to stabilise a high copper surface area and may also have a slight promotional effect on the copper. For the sample CuYZr 30/20/50, the level of deactivation was negligible. Each promoter stabilizes a higher total surface area, presumably by stabilizing the structure of the zirconia. The magnitude of the drop in surface area after operation for 10 h at 573 K is an indication of the all-over stability of the catalysts. The metal surface areas of the promoted samples were also higher and so it appears that the promoter also has some effect on the dispersion of the Cu. The values of the TOF's for all the promoted samples are higher than that of the unpromoted material, indicating that the active site must be affected in some way by the presence of the promoter and that the promoter does not just have the structural promotion effect shown by the total surface the promoter but also decorates the copper surface and therefore is likely also to affect the nature of the active site. Since dissolution of Y or La cations in the lattice of zirconia is well known to generate oxygen vacancies thus providing ionic conductivity as well as creating surface sites for water dissociation, it is possible that surface/bulk oxygen mobility in the complex oxide support could be also important for stability of the Cu-based catalyst performance. A correlation between the mobility of lattice oxygen in complex CeZrY -O based oxide support and activity/performance stability in steam reforming of methanol over $Cu_x(CeZrY)_{1-x}O_y$ catalysts (pure and with addition of Al_2O_3 and Cr) synthesized via the urea–nitrate combustion method was indeed observed [171]. The efficient transfer of lattice oxygen to copper clusters appears to facilitate transformation of reaction intermediates into desired products and prevent the formation of oxygenated polymers -precursors of coke (vide supra). Addition of alumina and Cr to $Cu_x(CeZrY)_{1-x}O_y$ catalysts leads to the steep increase in H_2 production with stabilization of catalytic activity. Alumina helps to disperse the active copper species, while chromium acts as a textural promoter of copper preventing it sintering.

In general, the components of bio-oil are numerous and the composition depends on the type of biomass and pyrolysis conditions, with its main components belong to the following groups: acids, aldehydes, alcohols, ketones, phenols, sugars and furans [58, 172, 173]. In most research work, some single light component derived from bio-oil was chosen as a model compound for catalyst testing, However, the overall reactivity of the fuel mixtures is not simply an average over the reactivities of constituent molecules. There has been very little research work on comparison of the steam reforming of different organic fractions derived from bio-oil. For such a complex organic mixture, it is necessary to investigate the reaction behaviors for steam reforming of different bio-oil fractions. In the work [20], two organic model compound mixtures were selected as two different representative fractions derived from bio-oil. First fraction was made up of methanol, ethanol, acetic acid and acetone in the equal weight proportion, which are the major classes of components present in the aqueous phase of bio-oil. The fraction with a low C/H ratio (0.250–0.500) is regarded as a light fraction representative. Second fraction was made of furfural, phenol, catechol and m-cresol, taken in the equal proportion as well, which are the common components in the organic phase of bio-oil. The fraction with a high C/H ratio (0.875–1.000) is regarded as a heavy fraction representative. This study has shown that there are significant differences in the gas product distribution and carbon deposition behavior from steam reforming of two different model bio-oil fractions. For the steam reforming of the light fraction, higher H_2 yield and carbon conversion to gas-phase components can be obtained at a relatively low temperature (650°C). However, for steam reforming of the heavy fraction, a higher temperature (800°C) is necessary to obtain higher H_2 yield and carbon conversion to gaseous products. For the efficient steam reforming the heavy fraction needs a higher steam-to carbon ratio (10) than the light fraction (7) at the chosen temperature of 800°C. Based on the same carbon space velocity, for 10 h time-on-stream, a drop of H_2 yield and carbon conversion into the gas phase products in the steam reforming of the heavy fraction was more rapid than that of the light fraction. Carbon deposition in the steam reforming of the heavy fraction is much more severe than that of the light fraction.

Most of publications about oil steam reforming and cracking is related to bio-oil (or pyrolysis oil) and tar derived from pyrolysis of biomass and wood, respectively. There are few scientific publications about gasification of mineral (lube) oil (C_{15}-C_{50}) [174, 175]. Catalytic reforming/cracking of mineral oil using effective Pt-supported doped Ce-Zr-O-based catalyst is not discussed in literature. Hence, experience of the team from Boreskov Institute of Catalysis in the oxy-steam reforming of mineral oil at short contact time on structured catalysts (vide infra) will be certainly of interest for specialists in the discussed subject.

2.3. THE TEXTURAL FEATURES OF STRUCTURED CATALYTIC SYSTEMS

These features are important for operation at high space-velocity with reducing mass and heat transfer resistances and, thus, maintaining a high overall throughput. In addition to the optimal catalyst active composition , a rational design of substrates is required. This macroscale level of catalyst design refers to the design of macrostructured catalytic materials in order to intensify the process as well as to improve the catalyst usability (easier separation, etc.) and resistance to deactivation. This includes a proper selection of material for substrate manufacturing and its geometry, since they both play a significant role in determining the effective thermal conductivity of the structured catalysts and reactors. Moreover, oxidation reforming chemistry is fast and often requires a low catalyst surface area to avoid deep oxidation [43]. Schmidt et al [23] studied platinum group metals supported extruded monoliths, foams and metal gauzes. Rh-containing catalysts ensure higher hydrocarbon conversions and syngas selectivities (up to 90%) as compared with Pt-containing ones, especially at high GHSV and pressures up to 5 atm [176-178]. For gauzes, oxygen conversion was incomplete and hydrogen selectivity was low [179]. Catalyst geometry rather than surface area is often responsible for syngas selectivity.

Effects of washcoat, pore size, type of ceramic foam substrate material and noble metal loading on CH_4 conversion and syngas selectivity were studied by Bodke et al. [26]. It was found that washcoat addition, decreasing foam cell size and replacement of zirconia for alumina increase syngas selectivity and reduce olefin selectivity irrespective of the fuel, catalyst or amount of diluents used. Most of these results can be explained on the basis of differences in the mass transfer rates.

A steep temperature profile along the length of monolithic catalyst layer is often present in reactors performing partial oxidation of hydrocarbons into syngas, because deep oxidation occurring in the inlet part is strongly exothermic whereas the reforming reactions are endothermic. Moreover, hot spots may also occur affecting both stability and safety of the process. The high temperature over the catalyst is a mayor threat to the stability of supported metal catalyst: the catalyst may sinter, lose activity and, most importantly, metals exposed to oxygen at these temperatures form volatile oxides resulting in metal loss. At the low heat conductivity of the catalyst bed, radial and axial temperature gradients may develop, leading to corresponding variations in both catalyst efficiency and selectivity, which are known to be very sensitive to temperature. It is explained by the fact that an increase in temperature favors the intrinsic kinetic constant with respect to diffusional constant. The influence of the intrinsic heat conductivity of the support material was considered in publications of Groppi et al [180-183] and others [184, 185]. The experiments with the plate-type catalyst with structured metallic supports show that heat conduction in the metallic supports of structured catalysts can provide an effective alternative mechanism to remove the heat generated by strongly exothermic reactions. The catalyst bed with high thermal conducting properties may act as a heat exchanger transferring heat from hot to cold zones.

The structured catalytic systems based on metallic supports are very attractive to perform endothermic reactions also due to their higher thermal conductivity and possibility of heating by the passing of an electric current through them [28, 184]. Several types of metallic substrates with different geometry comprised of different metals and alloys were successfully used at the Boreskov Institute of Catalysis to design structured catalysts and reactors for partial oxidation/steam reforming of hydrocarbons and oxygenates into syngas and hydrogen. Refractory alloys (fechraloy etc) as thin (50-100 micron thickness) foil or gauze (woven from the fecraloy wires with diameter 2.0 mm and ~0.2 mm spacing) were covered by strongly adhering non-porous layers of refractory oxides (stabilized alumina or zirconia) by using blast dusting technique [117, 122]. Monolithic substrates (diameter 55 mm, length 45 mm, specific surface up to 4700 m^2/m^3, porosity ~ 0.8) were obtained via stacking a flat and corrugated foils and winding them into an Arkhimed spiral. Standard washcoating procedures were used to deposit first a sublayer of γ-Al_2O_3, and then porous layers of complex perovskite-like or fluorite-like oxides. Active metals (Pt, Ru, Rh, Ni etc) were supported via wet impregnation. Typical images of these substrates are shown in Figure 9. After spark-welding of tungsten rods as electrical current leads, such a monolithic substrate was also used for evaporation of liquid fuels sprayed via a nozzle ensuring also a good mixing of fuel vapours with steam and air.

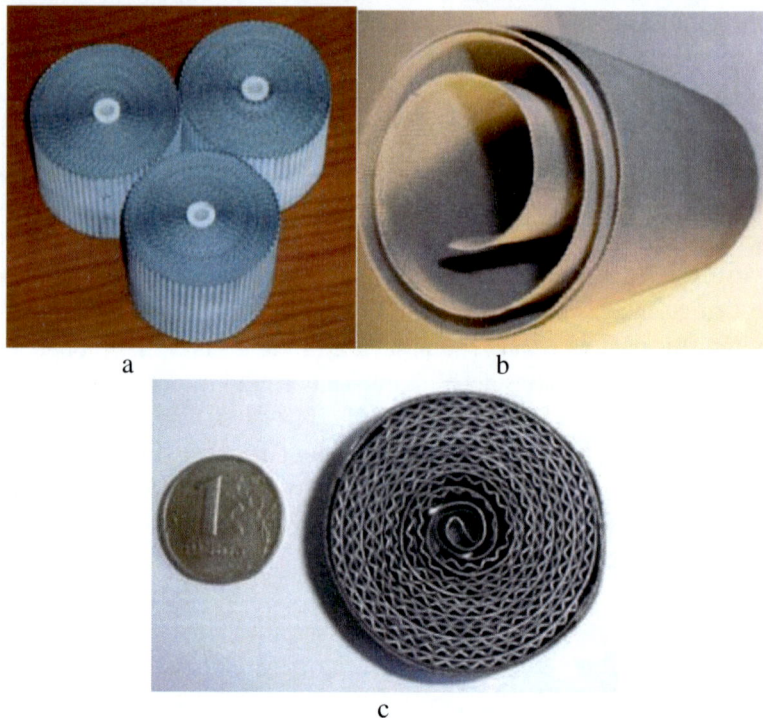

a   b

c

Figure 9. Typical images of Fechraloy substrates comprised of 20 microns foil (a), gauze (b) or 200 microns foil (c).

For design of a *radial flow structured catalytic reactor* (vide infra), a package consisted of 60 FeCrAl washers (outer diameter 38 mm, inner diameter 19 mm, height 2 mm) piled up and fixed into a cylinder by means of nut bolting was manufactured. Multiple radial channels (0.5 mm depth and 0.5 mm width) were milled in the washer planes with the angular step width of 4 degrees, resulting in 90 channels on a side. Protective layer of corundum was supported by blast dusting. The catalytically active $LaNi_{0.9}Pt_{0.1}O_3/CeZrO_x$ composition was supported on the protective layer using the same procedures as were applied for metallic foil monoliths.

For *catalytic microreactors* applied to study partial oxidation of natural gas and steam reforming of methanol (vide infra), porous/non-porous platelets of different shapes made of a variety of metals and alloys were used (Figure 10).

Active components were supported either by washcoating (for dense platelets) or by dry pressing of active components into the porous platelets.

Figure 10. Microchannel plates: cylindrical stainless steel base catalytic plate (a), cylindrical aluminum base catalytic plate (b), rectangular porous nickel based catalytic plate (c). Stainless steel microchannel plate manufactured by laser lithography with subsequent electrochemical etching (d).

Composite ceramometal monoliths (46 mm diameter, 23 mm length, ~ 300 cpsi channel density, ~ 0.3 porosity, Figure 11) were prepared by hydrothermal oxidation of blends comprised of powdered aluminum with fillers (powders of alloy, metals, other oxides, pore-forming agents etc) in specially designed dies [117, 186, 187]. To form a set of transport channels, easily burned organic fibers were inserted into the cermet matrix before the hydrothermal treatment stage. Parallel channels (~ 0.5 mm diameters) along the substrate monolith are interconnected by narrow (~0.1 mm) anfractuous channels separated by ~0.05 mm walls. Active components were loaded via the incipient wetness impregnation procedure.

More traditional ceramic monolithic substrates (a hexagonal prism with side of 28 mm, monolith length of 50 mm, wall thickness of 0.25 мм, equivalent channel diameter of 1.9 mm, and surface area of 3-10 m^2/g, Figure 12) were manufactured via extrusion of plastic pastes containing a binder and refractory oxides (corundum, partially stabilized zirconia, titanium aluminate etc). Active components comprised of Pt (Pt/LaNiO$_3$) -promoted Ce-Zr-La-O mixed oxides were supporting via successive impregnation procedures [123].

Figure 11. Typical image of composite porous ceramometal monolithic substrate.

Figure 12. Typical image of extruded ceramic thin wall honeycomb monolithic substrates.

In this work we have reviewed some representative results obtained at the Boreskov Institute of Catalysis and presented in details in the original publications [117-119, 123, 188-192] in the study of synthesis gas formation from hydrocarbon fuels and oxygenates by using both standard and new structured catalytic reactor configurations.

Chapter 3

THERMODYNAMICS OF FUELS REFORMING INTO SYNGAS

At full conversion of a carbon/hydrogen/oxygen system, irrespective of the mode of reactors operation, the composition of the exit gas will be determined by thermodynamic equilibrium corresponding to the outlet temperature, which is controlled by the inlet temperature, the feed composition, the reaction heat evolved and heat exchange with the environment. Obviously, this is only the case if all necessary reactions can take place because the proper catalyst is used, and the residence times in the catalyst are long enough [193]. By using thermodynamic study, a better correlation can be established between the product composition, temperature, overall energy conversion efficiency, and different operating conditions such as the oxygen-to-carbon ratio and/or the steam- to-carbon ratio. Therefore, there are thermodynamically optimal reaction conditions, which are of high practical relevance for each chemically reversible reaction.

Calculating the composition in a carbon/hydrogen/oxygen system at chemical equilibrium can be done by either specification of possible reactions taking place or minimizing the total Gibbs free energy in the given chemical system. The equilibrium describes the temperature of the system and the product composition. Generally, the thermodynamic equilibrium in a reformer reactor depends on the following parameters:

- preheat temperatures of the reactor feed (air, water, and fuel)
- pressure inside the reactor
- chemical composition of the fuel
- heat loss
- fuel-to-air and/or fuel-to-water ratios.

Fuel/air/steam ratio in the feed is a primary parameter which determines both hydrogen yield in the product gas (reformate) and the energy released or absorbed by the reaction and, hence, the adiabatic reaction temperature. A correct representation of the thermochemical properties of hydrocarbon fuel is very important for providing a more realistic estimation of expected parameters. This requires identification of possible components. However, very often hydrocarbon fuels and oxygenates are complex mixtures of different compounds. Thus, gasoline fuel is a mixture of normal paraffins, isoparaffins, cycloalkanes, aromatics and olefins.

Let us demonstrate thermodynamic equilibrium restrictions on the performance of syngas formation from methane, isooctane, gasoline and n-decane. We have tested gasoline containing 191 types of hydrocarbons. The amount of aromatics was about 40 wt%. According to the chromatographic analysis data, a generic formula for the averaged composition corresponds to $C_{7.2}H_{13.36}$. In the case like that, a model fuel is required which is sufficiently flexible to incorporate a large number of compounds/interactions and provide good results [194-197].

The surrogate gasoline mixture that was specified for modelling was an equilibrium distribution of 29 close-cut C3-C10 fractions identified according to the following three criteria: sum of carbon and hydrogen atoms identical to that of gasoline; closely matching fractional distillation diagrams; and low heating value (Table 1).

The characteristics of the fuels, both of the base gasoline employed in the experimental study and the surrogate used as a model in the thermodynamic calculations, are summarized in Figure 13 and Table 2. As anticipated, the two types of fuels display good similarity. Available data from the experimental observations confirmed the reliability of such approach: both the product composition and the catalyst outlet temperature were very close to the values thermodynamically predicted by using the model of the base gasoline.

Specification of the system components (not the reactions) was used in the thermodynamic consideration. The equilibrium model was based on the following simplifications: uniform temperature and pressure are assumed; no information about actual reaction pathways/formation of intermediates; no tar, no solid carbon are modeled; no information about the rate of the reactions. In addition to the reacting system (a mixture of air, steam, and organic compounds as methane, isooctane and surrogate gasoline), such components as H_2, CO, CO_2, H_2O, CH_4, C_2H_2, C_2H_4, C_2H_6, C_3H_6, C_3H_8 were taken into account to determine the resultant mixture composition as a function of oxygen-to-carbon ratio, steam-to-carbon ratio, temperature, pressure, and product species. The Peng–Robinson equations

of state were used to calculate the stream properties with the HYSYS software
package.

Table 1. Model fuel mixture exhibiting similar characteristics to gasoline

N	Component	C	H	Mass fraction, w/w %	$\Delta_f H°$, kcal/mol
1	Isobutane	4	10	0.430	-37.87
2	n-Butane	4	10	0.970	-35.29
3	Isopentane	5	12	1.540	-42.95
4	n-Pentane	5	12	1.260	-41.40
5	2,2-Dimethylbutane	6	14	0.320	-51.00
6	2,3-Dimethylbutane	6	14	0.320	-49.48
7	2-Methylpentane	6	14	1.130	-48.82
8	3-Methylpentane	6	14	4.830	-48.28
9	Hexane	6	14	3.700	-47.52
10	Methylcyclopentane	6	12	0.820	-33.08
11	cyclogexane	6	12	1.180	-37.34
12	2,2-Methylpentane	7	16	6.319	-57.05
13	2,3-Methylpentane	7	16	4.140	-55.81
14	3 -Metylhexane	7	16	7.189	-54.35
15	Heptane	7	16	4.140	-53.63
16	Isooctahe	8	18	0.270	-61.97
17	2,3+2,2 -Dimethylhexane	8	18	1.590	-60.40
18	2,3,4 -Trimethylpentane	8	18	0.270	-60.98
19	2,3,3 -Trimethylpentane	8	18	0.270	-60.63
20	2 -Methylheptane	8	18	22.258	-60.98
21	Octane	8	18	1.860	-59.74
22	Benzene	6	6	10.209	11.72
23	Toluene	7	8	14.429	2.87
24	m-Xylene	8	10	4.220	-6.07
25	p- Xylene	8	10	1.760	-5.84
26	o- Xylene	8	10	1.760	-5.84
27	Ethylbenzene	8	10	0.700	-2.98
28	p+o+m-Diethylbenzene	10	14	1.060	-17.44
29	1,2,4 -Trimethylbenzene -	9	12	1.060	-14.78

Table 2. Characteristics of the base gasoline and surrogate mixture

Fuel characteristic	Base gasoline	Model gasoline
C/H ratio	0. 539	0.539
Low heating value, MJ/kg	44.5	43.1
Data of fractional distillation, °C :		
Start point	42	38
10 wt. %	73	71
50 wt. %	112	110
90 wt. %	165	170
End	185	182
Density, g/ml (20°C)	0.78	0.8

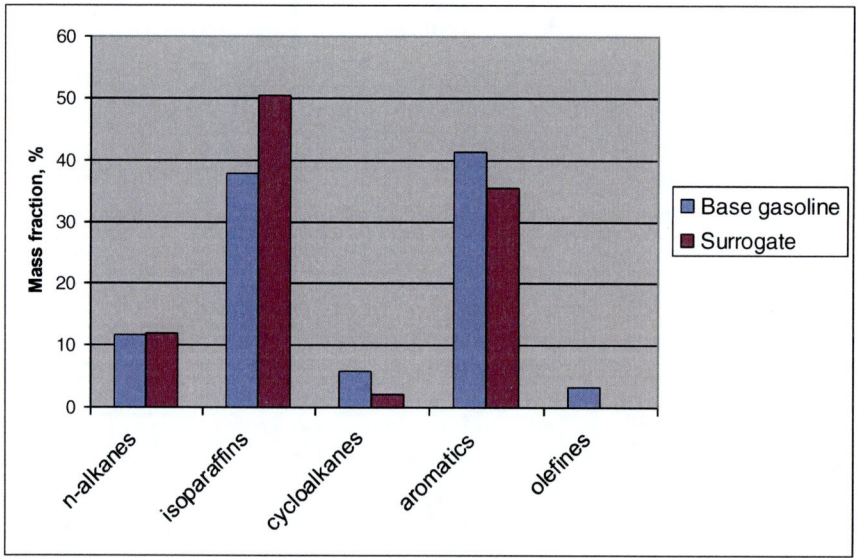

Figure 13. Fractional composition of the base gasoline and its surrogate mixture.

It is desirable for reformed gas to have H_2 content as high as possible. Obviously, the concentration of hydrogen in the product gas will depend on the number of hydrogen atoms per carbon in the fuel. The maximum achievable concentration of hydrogen for the fuels considered decreases in the order: methane (41%)> isooctane (28.1%)> gasoline (24.3%), according to the reaction stoichiometry. The actual product composition and efficiency of the reforming process depend also on reactivity of the specific fuel and its components. It is clear that the efficiency of fuel reforming into synthesis gas is directly proportional to the amount of hydrogen and carbon monoxide generated and reaches its maximum at the minimum value of the air-to-fuel ratio, and its minimum -at the maximum of the air-to-fuel ratio. The syngas formation efficiency is also coupled to both the feed and reaction temperature [196].

3.1. EFFECT OF THE REACTION TEMPERATURE

Figure 14 shows the product distribution in the reformed gas versus reaction temperature for a given O_2/C molar ratio in the given feed. It can be seen that for all hydrocarbon fuels considered in the calculations, the conversion does not require too high temperatures. In a practical temperature range of $800 - 1200°C$ for the partial oxidation reforming reaction occurring at 1 atm, concentrations of hydrogen and carbon monoxide in the product gas change only weakly. The point is to discover an optimal value of the reaction temperature required, e.g., to reduce the carbon deposition and methane formation which are thermodynamically favored at lower temperatures. Yet, the higher operational temperatures during partial oxidation/reforming would tend to formation of by-products (e.g. olefins, cyclic products).

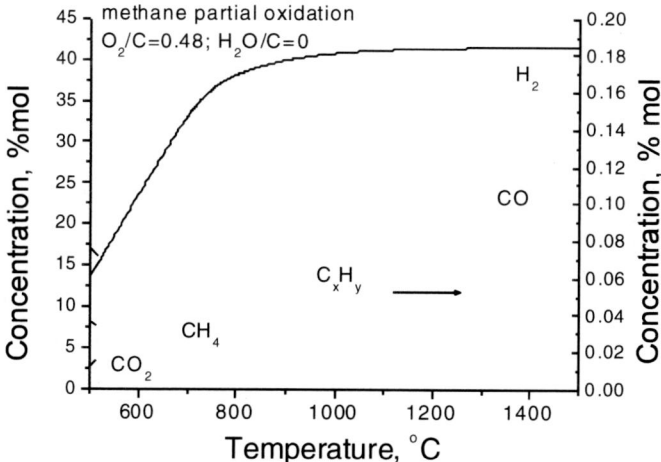

Figure 14. Continued on next page.

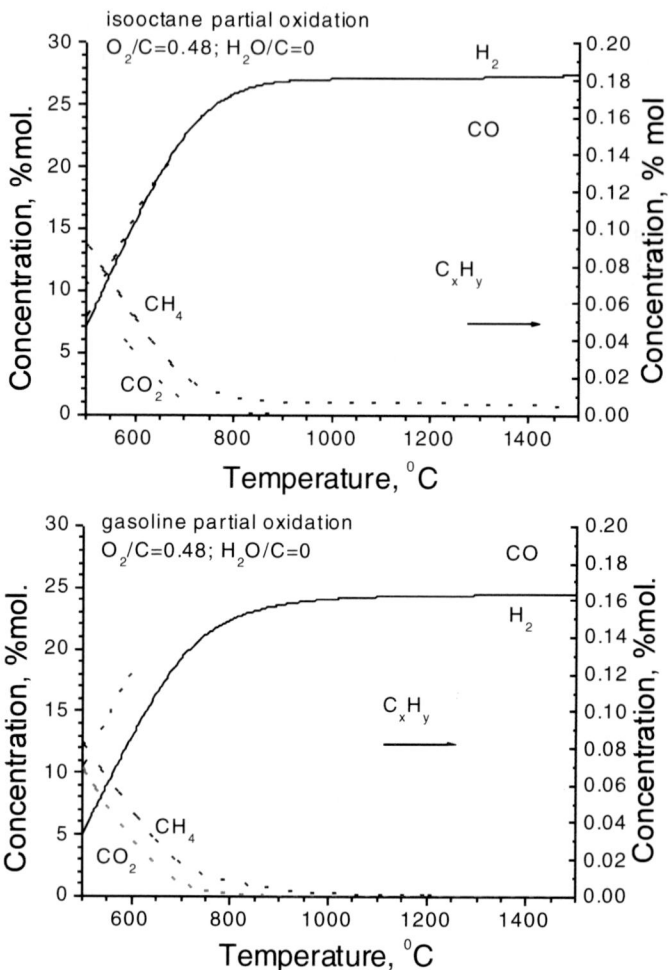

Figure 14. Thermodynamic predictions of equilibrium gas composition produced in the isothermal partial oxidation of methane, isooctane and gasoline with air as a function of temperature. P=1 atm.

The thermodynamic limits for the case of decane selective oxidation in the isothermal reactor are shown in Figure 15. In the range of 800 - 1200 ^{0}C, which is a practical temperature range of the process, concentrations of syngas key components - hydrogen and carbon monoxide change only slightly. However, for the all range of the temperatures considered the formation of the secondary hydrocarbons is favorable, indicating by this a severe tendency to the coke formation.

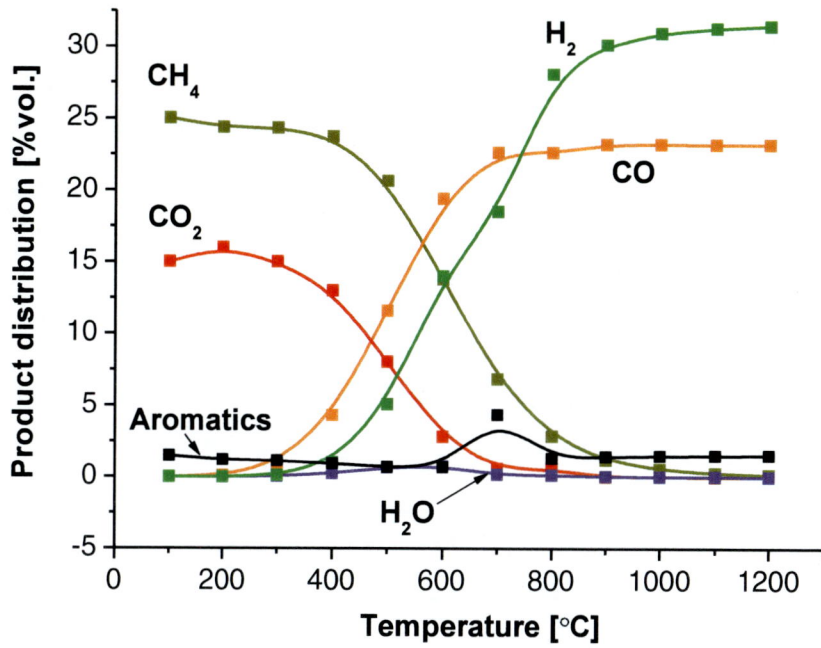

Figure 15. Effect of the temperature on the product distribution for n-decane partial oxidation in the isothermal reactor.

3.2. EFFECT OF PRESSURE

The chemical engineer would like to increase the extent of reaction and, as a result, the yield of useful products. However, from Le Chatelier´s principle, which is a consequence of the Second Law of thermodynamics, increasing the extent of a reaction which proceeds with the increase in the number of moles, requires a reduction in the pressure.

This pressure variation decreases the Gibbs energy of reaction (increases driving force), thus driving the reactions further towards completion. Similarly, the Le Chatelier principle requires a reduction in the temperature of exothermic reactions (and an increase in the temperature of endothermic reactions).

Thus, the increase of pressure has a negative effect on both the hydrogen yield and the operational temperature (see Figure 16). For gasoline partial oxidation the lower limit of about 750°C could be determined at 1 atm, while the pressure of 15 atm results in the operational temperature of about 1000°C.

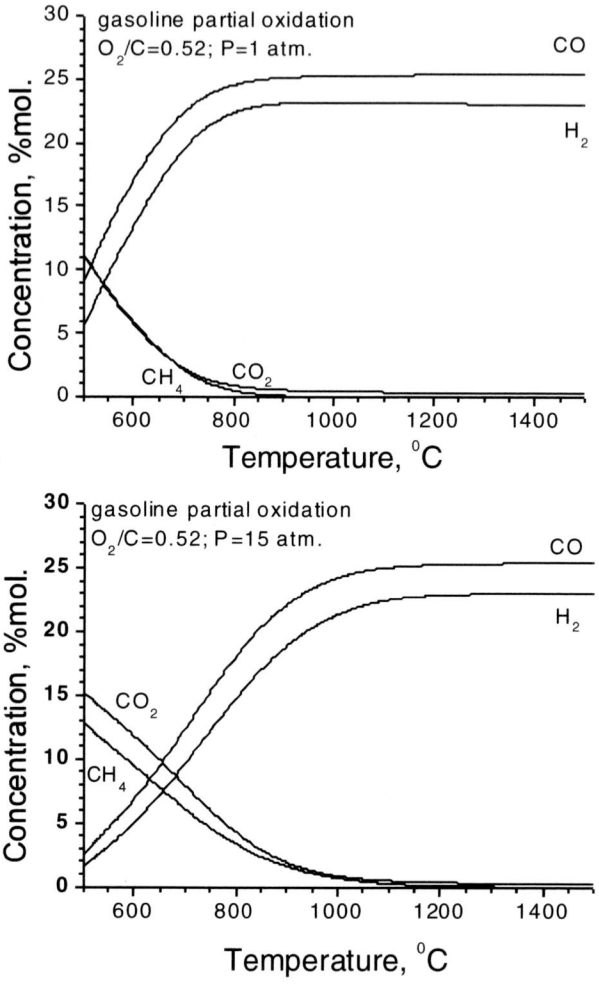

Figure 16. Effects of the temperature and pressure on main products in the equilibrium reformate gas at the isothermal partial oxidation of gasoline with air.

3.3. EFFECT OF THE FEED PREHEATING TEMPERATURE

Obviously, the higher preheat temperature results in the lower reforming efficiency. Effect of the feed temperature on the reaction temperature in adiabatic reactor and syngas formation from methane and gasoline may be clarified by the thermodynamic considerations shown in Figure 17.

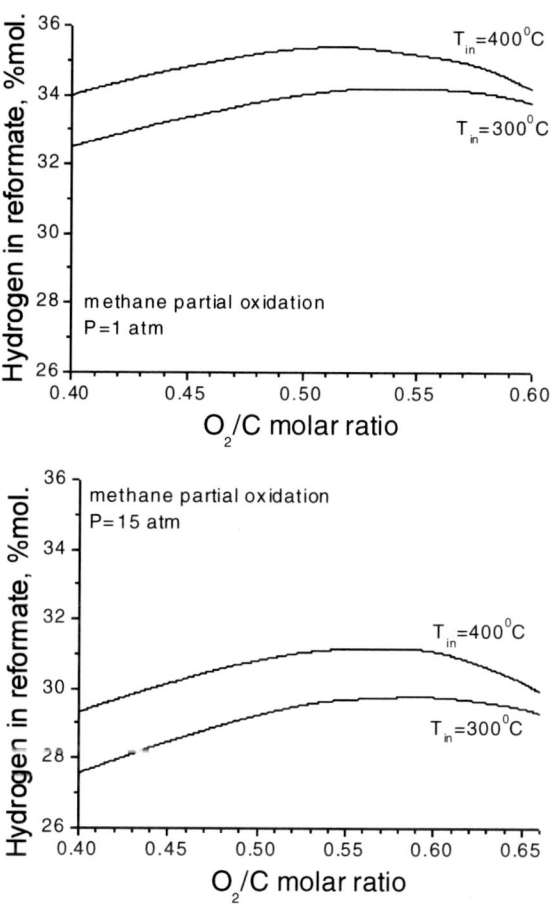

Figure 17. Effect of the O_2/C ratio in the methane-air feed with the different preheat temperature on the hydrogen concentration in the product gas (adiabatic reactor). Thermodynamic predictions for P=1atm (left) and P=15 atm (right).

At designing a fuel preheating system including the fuel-air mixture preparation with fuel atomization and vaporization, one should take into account the fuel reactivity towards thermal cracking and pre- reforming reactions in the gas feed. Although methane is the most abundant of alkanes, it is also the least reactive [198].

The preheat temperature of the methane-air mixture can exert an important effect on the choice of the operational O_2/C ratio (see Figure 17), the reaction temperature, and syngas formation. Similar trends were observed for gasoline partial oxidation (Figure 18).

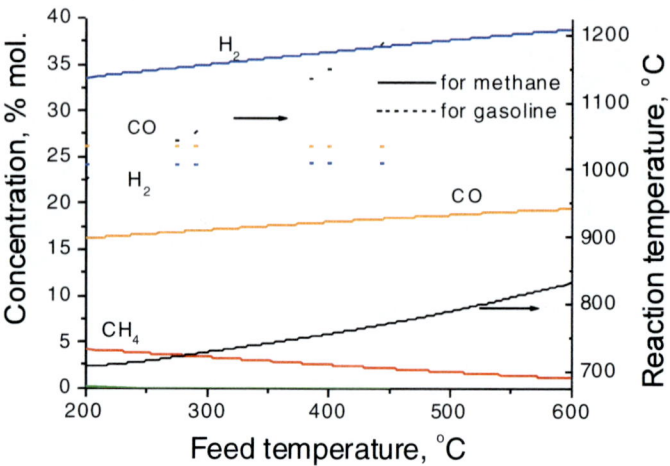

Figure 18. Thermodynamic predictions on the effect of the feed temperature in the partial oxidation of methane and gasoline by air at adiabatic conditions. $O_2/C=0.50$, P= 1.0 bar.

The experimental data collected over honeycomb corundum catalysts (Figure 19) were also consistent with the stated thermodynamic predictions for the methane partial oxidation.

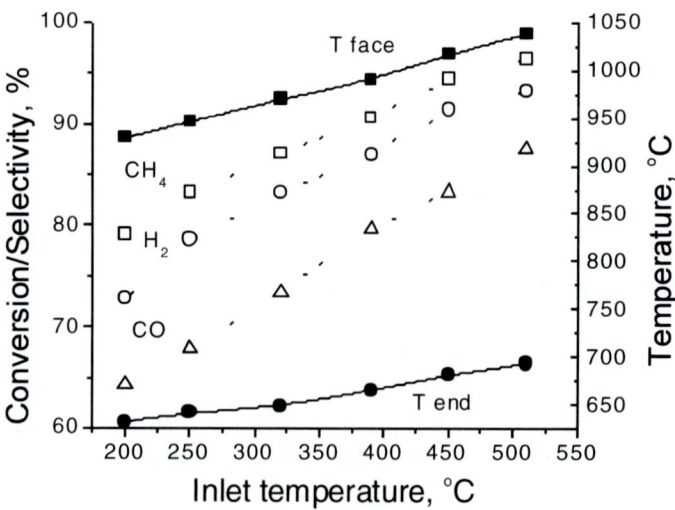

Figure 19. Experimental data on the effect of inlet temperature on the process performance over corundum monolith (0.2% Pt/ 7% $LaNiO_3$) with triangular shape channels (d_{eq} =0.67 mm, ε=0.6): 27 % CH_4 in air, U_o=0.4 m/s (STP), 0.15 s (STP) contact time.

However, isooctane and even more so gasoline fuels are much more reactive towards cracking compared to methane, especially in the mixture with air. According to direct experimental observations in the isooctane partial oxidation, the catalyst temperature tends to decrease with preheating of the isooctane-air mixture (Figure 20, left).

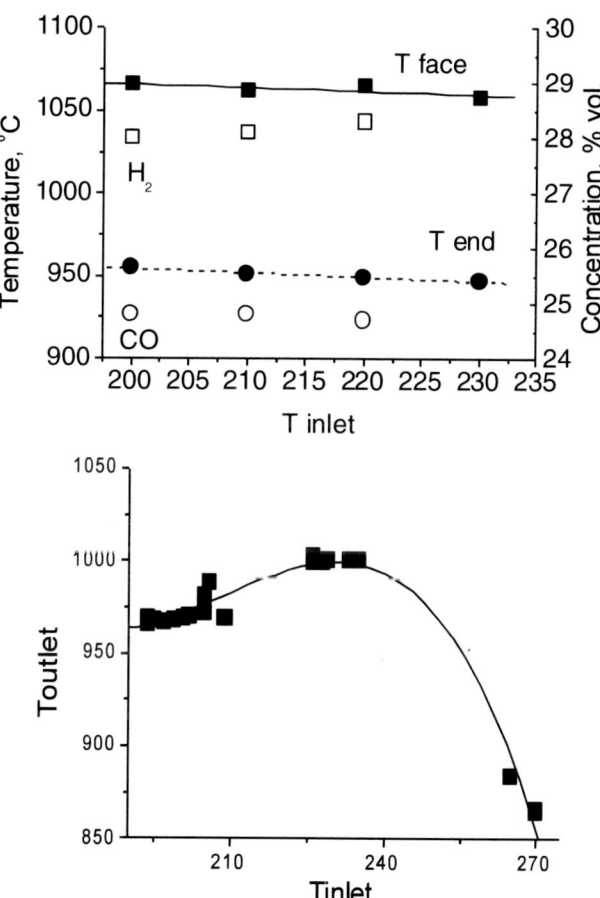

Figure 20. Effect of the inlet temperature on the partial oxidation process performance over monolithic catalyst on thin-foil fechraloy substrate. Left: isooctane 0.757 kg/h in air, $O_2/C=0.53$, 0.1 s (STP) contact time; Right: gasoline 0.894 kg/h in air, $O_2/C=0.53$, 0.1 s (STP) contact time.

The fractional composition of gasoline was observed to be altered even at 200°C preheat of the feed [190, 191]. In the gasoline partial oxidation, the catalyst temperature is just slightly increased with an increase in the temperature of gasoline-air mixture to about 240°C and then decreased with further preheat due to non-catalytic pre-reforming reactions (Figure 20, right). With a due regard for this fact, the preheat temperature of about 190°C was selected to be the preferable one for the gasoline-air feed.

The effect of the preheat temperature of n-decane on the temperature in the adiabatic reactor and product gas is shown in Figure 21.

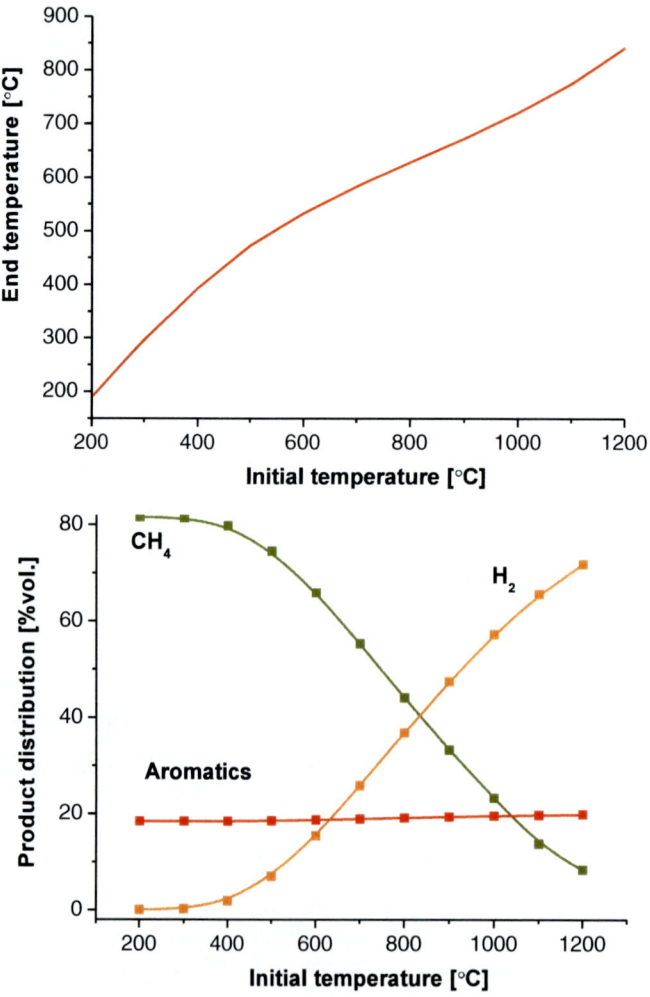

Figure 21. Dependence of the temperature (left) and the product distribution (right) on the initial temperature at the reforming of n-decane.

It can be seen that for adiabatic reactor, there is no strong effect of the preheat (initial) temperature on the temperature of the product gas. Even at the initial temperature of 1200°C, the temperature of the product mixture could be only 358 degrees less. Amount of hydrogen in the product gas would increase considerably with the preheat temperature. Obviously, this is only the case if all necessary reactions can take place because the proper catalyst is used, and the residence times in the catalyst are long enough. However, the presence of aromatics in the product gas clearly suggests that coke formation is a serious problem at the preheating of n-decane.

3.4. EFFECT OF THE O_2/C MOLAR RATIO

Effect of the O_2/C molar ratio in the feed on the level of hydrogen in the product gas and the reaction temperature is found to be much more pronounced for gasoline reforming than those for methane and even for isooctane (Figure 22). Optimal processing of gasoline occurs in the restricted range of O_2/C =0.50-0.52 in which the concentrations of hydrogen in the product gas correspond to the calculated values of 24.08-23.15 % mol, respectively (dry gas). For the fixed input temperature (200°C), thermodynamic calculations show that with variation in the O_2/C molar ratio from 0.5 to 0.52 in the feed the adiabatic temperature rise increased from 787 to 840 degrees. Further increasing the O_2/C molar ratio may cause deactivation of the catalyst due to the high temperature in the front part of the monolith, as it was observed by Bobrova et al [189]. Compared to gasoline, the moderate heat evolved and smaller increase in the temperature (about 40 degrees) is observed for isooctane in the same range of the O_2/C ratio. This enables operating at higher O_2/C ratios in the isooctane-air mixture. Much weaker dependence on the O_2/C ratio is observed for methane partial oxidation. In this case, it is desirable not only to preheat the feed mixture to a high temperature, but operate at O_2/C ratio of about 0.6, which is higher than the stoichiometric one (see also Ref. [199]).

Effect of both air-fuel ratio expressed as O_2/C molar ratio in the feed and preheat temperature for partial oxidation of n-decane may be clarified by consideration of calculated data shown in Figure 23. The outlet temperature in these Figures represents the operational temperature in the adiabatic reactor at given inlet conditions. In the range of O_2/C ratio considered, the adiabatic temperature strongly depends on the air-fuel ratio in the feed likewise for gasoline. At increasing O_2/C ratio above 0.5, both hydrogen and carbon monoxide amount in the product gas strongly declines because the excess of oxygen in the feed oxidizes H_2 and CO to H_2O and carbon dioxide resulting in the operational temperature increase in the adiabatic reactor.

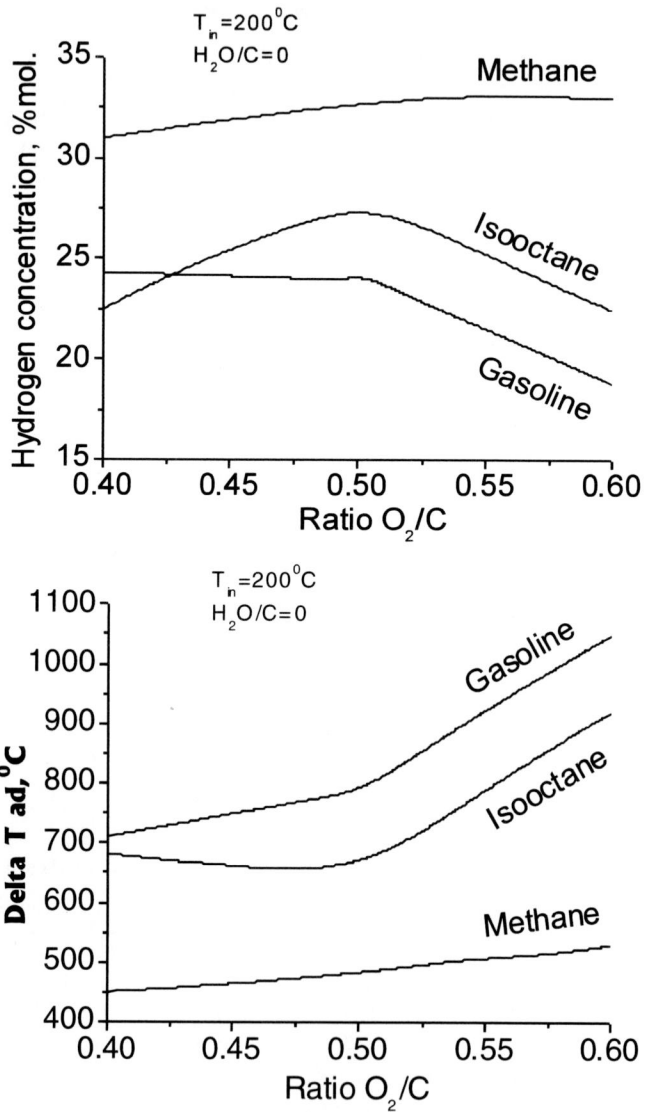

Figure 22. Effect of the molar oxygen–to-carbon ratio in the fuel-air mixture on the hydrogen concentration and adiabatic temperature rise at the partial oxidation of methane, isooctane and gasoline with air.

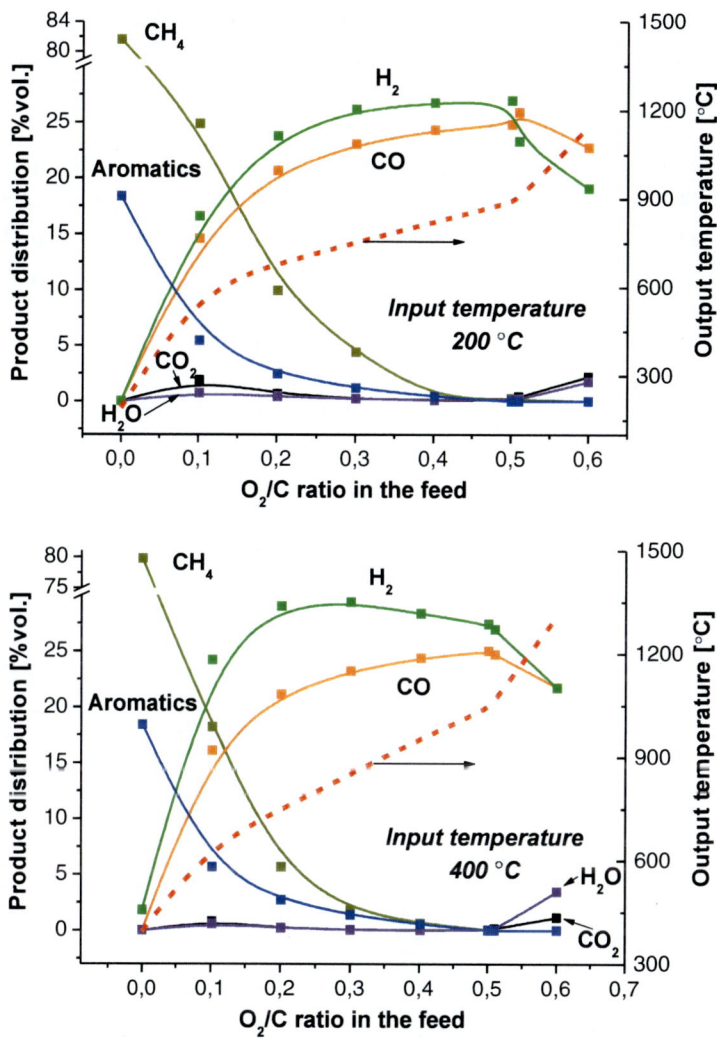

Figure 23. Thermodynamic predictions on the effect of the air- fuel ratio expressed as O_2/C molar ratio in the feed on the product distribution and the temperature in adiabatic reactor at $T_{in}=200°C$ (left) and $T_{in}=400°C$ (right) at the partial oxidation of n-decane.

The preheat temperature can exert an important effect on both the adiabatic temperature and product distribution due to a high exothermicity of the reaction. One can see that in terms of the hydrogen content, an optimum value for the O_2/C ratio is 0.5 at the input temperature of 200°C. This value becomes 0.3 at the input temperature of 400°C. Again, one should take into account the reactivity of a given fuel towards cracking, especially in the mixture with air. Our experiments demonstrated [190] that the predictions for the equilibrium composition at the monolith back face temperature in the autothermal reforming mode (with steam addition) are in very good agreement with the experimental data as well. Effects of such parameters of the feed gas as the steam and oxygen-to-carbon ratios on the autothermal reforming of gasoline are shown in Figure 24. It is important to feed sufficient oxygen, so the energy generated by the oxidation compensates the energy absorbed by the endothermic reactions.

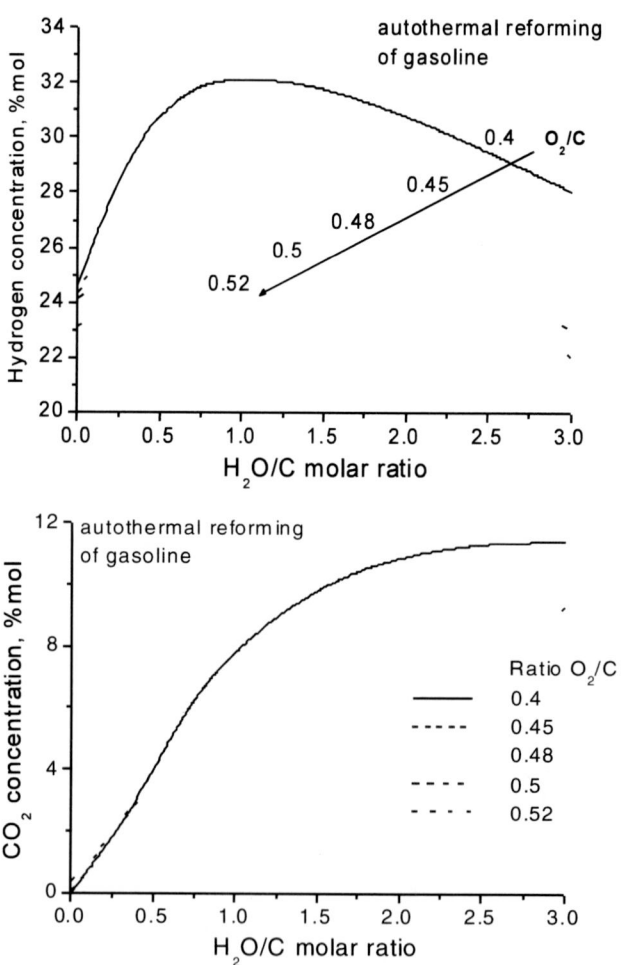

Figure 24. Effect of the H_2/C and O_2/C molar ratios on the content of H_2 and CO_2 in the product dry gas in the autothermal reforming of gasoline. Feed temperature $T_{in}=270°C$

Well defined maximum in the hydrogen concentration in dry reformate is observed at the H_2O/C molar ratio of about 1.0 in the considered range of O_2/C ratio in the feed (Figure 24, left). This value will correspond to the maximum reformer efficiency, because further increase in the hydrogen yield would require an additional energy input to dry the reformate gas. To keep the reaction temperature at about 800°C for the given input temperature of 270°C and steam-to carbon ratio of about 1.0, the oxygen-to-carbon ratio needs to be maintained at about 0.45. For these operational parameters, the thermodynamic calculations predict the hydrogen concentration in the product gas of about 29% mol.

3.5. REFORMING OF BIOMASS-DERIVED PRODUCTS

Condensable vapors (i.e. bio-oil) derived from pyrolysis of biomass are very complex mixtures of different compounds. Thermodynamic calculations of reforming of particular model compounds of bio-oil are usually performed to elucidate the optimum operational conditions. Thus, such compounds as acetic acid, ethylene glycol and acetone were used in Ref. [199-201] in order to define the optimum conditions for maximizing the hydrogen production. It was found that the bio-oil oxygenates in the presence of steam are easily converted to hydrogen- rich mixtures achieving maximum hydrogen yield (80–90%) at 625°C without any coke formation for operation under atmospheric pressure and steam-to-carbon ratio >1. Thermodynamic analysis of ethanol processor systems [202] has shown that the most important parameter which affects efficiency in hydrogen production is the water to ethanol molar ratio in the feed. The values higher than stoichiometry result in the reduced efficiency because of increased enthalpy needs for water evaporation.

The thermodynamics of steam assisted high-pressure conversions of model components of bio-oil – isopropyl alcohol, lactic acid and phenol – to synthesis gas (H_2 + CO) have been investigated in [203] to understand the effects of process variables such as temperature and inlet steam-to-fuel ratio on the product distribution. It proved possible to adjust the H_2/CO ratios and the amounts of CH_4 and CO_2 in synthesis gas by changing the steam-to-fuel ratio, the value depending on temperature and the fuel type.

Thus, by using the thermodynamic calculations we can establish a correlation between the inlet and outlet characteristics of an adiabatic reactor at steady state. Complex interaction between the reactive flow and reactions on the catalytic surface occurs in the structured catalytic layers of different forms such as foam or extruded monoliths, wire gauzes, or sintered spheres, in a reforming reactor. Quite often the catalytic processes which can be run nearly autothermally and adiabatically exhibit an extremely fast variation of temperature, velocity, and transport coefficients of the reactive mixture, especially near the catalyst entrance. Spatio-temporal profiles developed in the reactors are results of interplay among kinetics, hydrodynamics, and the heat transfer. Generally, for a practical consideration, the operational characteristics should be chosen on the base of analysis of many factors including a range of throughputs, design of the reactor and configuration of monoliths, as well as the catalytic activity. Other issues such as catalyst kinetics or safety aspects appear to play an important role in determining the operating conditions of the reforming reactor.

The understanding of the details of the reactor behavior demands a better agreement between experimental and modeled flow field without neglecting the complex chemistry. Quite often, a *direct* experimental investigation of a reaction mechanism is difficult if possible at all under realistic conditions. Simulation studies are very helpful in examining physical-chemical processes in the reactor in detail. To achieve this, elementary-step kinetic models should be coupled with appropriate reactor models, and the simulation results should then be tested against available experimental data such as conversions and selectivities for the specific reactor configurations. Particularly, transient experiments and dynamic simulations, such as ignition and extinction of reactions, can often yield important insights into reaction mechanisms also due to their high sensitivity to the specific reaction path taken during the transient excursion of the reaction system [204-210].

In the next section the details of reaction behavior inside monolithic catalyst has been examined by mathematical modeling of the methane partial oxidation on the base of detailed chemistry.

Chapter 4

NUMERICAL STUDY OF THE PARTIAL OXIDATION OF METHANE

The basic reactor and methane partial oxidation reaction model for the Pt/Ce-Zr-La-O/α-Al$_2$O$_3$ honeycomb monolith has been presented in detail in a separate paper [211]. A dynamic one-dimensional two-phase reactor model of the processes with accounting for both transport limitations in the boundary layer of a fluid near the catalyst surface and detailed molecular unsteady-state kinetic model for surface reactions has been developed and verified with the transient experiments data. A mechanism proposed by P. Aghalayam et al. [212] was taken as a basis for the unsteady-state kinetic model to be implemented in the dynamic reactor model after some modification. The Pt/Ce-Zr-(La)-O/α-Al$_2$O$_3$ honeycomb monolith is a complex catalytic system where washcoat (fluorite-like nano-crystallites of solid solution Ce-Zr-La-O) strongly interacts with the active component -Pt. Study of detailed kinetics and mechanism of the partial oxidation reaction using the step response technique has made it clear that the active oxygen of the surface or bulk of complex oxide support quickly re-oxidizes the reduced platinum [116]. The simplified reaction scheme illustrates the reaction mechanism over Pt/CeO$_2$-ZrO$_2$ catalyst:

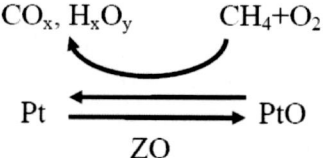

The unsteady state kinetic model of the partial oxidation of methane over the catalyst account for the interaction of washcoat (Z, ZO) with the active catalyst sites (Pt, PtO) (reaction 32 in Table 3). The implemented detailed mechanism contains 32 elementary steps of methane oxidation, 14 gaseous compounds and 13 intermediates on the catalyst surface with corresponding kinetic parameters.

The reactor model coupled with the detailed elementary step unsteady-state kinetic model was applied to test the simulation results against available transient experimental data of the reaction ignition during start-up. Due to their high sensitivity to the specific reaction path taken during the transient excursion of the reaction system, good correlations in such comparison approve reliability of the applied mathematical model [213, 214].

The experimental data with the full-size hexagonal corundum extruded monoliths with triangular shape channels (Figure 12) and active component comprised of 0.4 wt.% Pt/10 wt.% La-Ce-Zr-O were taken for the comparison with the numerical results. The numerically predicted gas-phase axial temperature profiles as the functions of time (curves) in comparison with the measurements (symbols) are given in Figure 25.

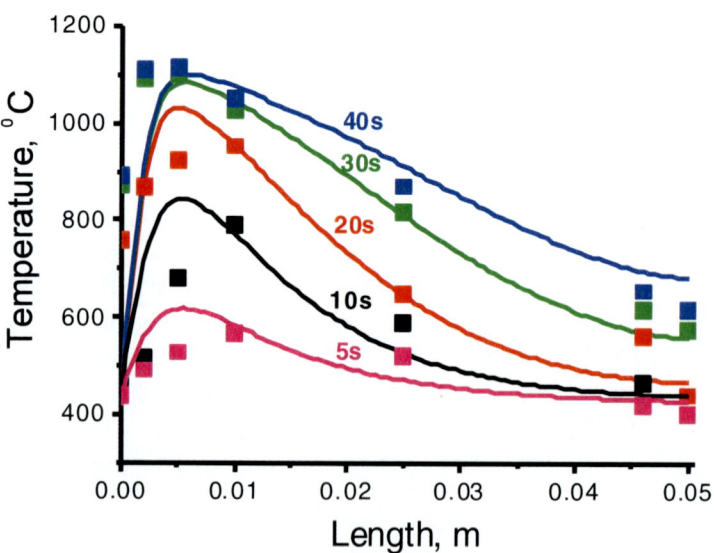

Figure 25. Gas temperature evolution in extruded monoliths with triangular shape channels during start-up. Feed mixture: 24 % CH_4 in air, T=420°C, P=1atm, u_o=0.5 m/s (STP). Lines - modeling predictions; symbols - experimental data. Adapted with permission from Ref 211. Copyright 2007 Elsevier B.V.

Table 3. Surface reaction mechanism for oxidation of methane over Pt/Ce-Zr-La/α-Al$_2$O$_3$ [211]

No	Reaction	k_f Preexponential, s^{-1} or sticking coeff.	E_f, kcal/mole	k_b Preexponential, s^{-1}	E_b, kcal/mole	$\Delta_r H$, kcal/mole
1	$OH^* + ^* \leftrightarrow H^* + O^*$	5.60×10^{11}	18.3	1.70×10^{10}	13.4	4.9
2	$H_2O^* + ^* \leftrightarrow H^* + OH^*$	1.20×10^{10}	39.1	3.50×10^{11}	0.0	39.4
3	$H_2O^* + O^* \leftrightarrow 2OH^*$	1.00×10^{11}	34.1	1.00×10^{11}	0.0	34.1
4	$H_2 + 2^* \leftrightarrow 2H^*$	0.09	0.0	3.33×10^{12}	20.0	-16.0
5	$O_2 + 2^* \leftrightarrow 2O^*$	0.03	0.0	1.00×10^{11}	19.0	-15.0
6	$H_2O + ^* \leftrightarrow H_2O^*$	1.00	0.0	5.33×10^{12}	10.0	-10.0
7	$OH + ^* \leftrightarrow OH^*$	1.00	0.0	1.00×10^{13}	30.0	-30.0
8	$H + ^* \leftrightarrow H^*$	1.00	0.0	1.00×10^{13}	60.2	-60.2
9	$O + ^* \leftrightarrow O^*$	1.00	0.0	1.00×10^{13}	67.0	-67.0
10	$CH_4 + 2^* \leftrightarrow CH_3^* + H^*$	0.68	12.0	3.97×10^{10}	5.5	6.5
11	$CH_3^* + ^* \leftrightarrow CH_2^* + H^*$	1.32×10^{13}	25.8	4.04×10^{10}	6.1	19.7
12	$CH_2^* + ^* \leftrightarrow CH^* + H^*$	1.00×10^{11}	25.0	1.00×10^{11}	12.2	12.8
13	$CH^* + ^* \leftrightarrow C^* + H^*$	1.00×10^{11}	5.4	1.00×10^{11}	37.6	-31.4

Table 3. (Continued)

No	Reaction	k_f Preexponential, s⁻¹ or sticking coeff.	E_f, kcal/mole	k_b Preexponential, s⁻¹	E_b, kcal/mole	$\Delta_r H$, kcal/mole
14	$CH_3^* + O^* \leftrightarrow CH_2^* + OH^*$	1.00×10^{11}	17.7	1.00×10^{11}	3.1	14.6
15	$CH^* + OH^* \leftrightarrow CH_2^* + O^*$	1.00×10^{11}	13.2	1.00×10^{11}	20.5	-7.3
16	$C^* + OH^* \leftrightarrow CH^* + O^*$	1.00×10^{11}	38.2	1.00×10^{11}	1.5	36.7
17	$CH_2^* + H_2O^* \leftrightarrow CH_3^* + OH^*$	1.00×10^{11}	19.5	1.00×10^{11}	0.0	19.5
18	$CH^* + H_2O^* \leftrightarrow CH_2^* + OH^*$	1.00×10^{11}	26.7	1.00×10^{11}	0.0	26.7
19	$C^* + H_2O^* \leftrightarrow CH^* + OH^*$	1.00×10^{11}	70.9	1.00×10^{11}	0.0	70.9
20	$CO^* + {}^* \leftrightarrow C^* + O^*$	1.00×10^{11}	74.2	1.00×10^{11}	0.0	75.
21	$CO_2^* + {}^* \leftrightarrow CO^* + O^*$	1.00×10^{11}	43.1	1.00×10^{11}	0.0	43.1
22	$CO + {}^* \leftrightarrow CO^*$	0.71	0.0	1.21×10^{13}	34.0	-34.0
23	$CO_2 + {}^* \leftrightarrow CO_2^*$	0.7	0.0	1.46×10^{12}	17.0	-17.0
24	$CO_2^* + H^* \leftrightarrow CO^* + OH^*$	1.00×10^{11}	38.2	1.00×10^{11}	0.0	38.2
25	$CO^* + H^* \leftrightarrow CH^* + O^*$	1.00×10^{11}	106.0	1.00×10^{11}	0.0	106.0
26	$CO^* + H^* \leftrightarrow C^* + OH^*$	1.00×10^{11}	69.2	1.00×10^{11}	0.0	69.2

No	Reaction	k_f Preexponential, s^{-1} or sticking coeff.	E_f, kcal/mole	k_b Preexponential, s^{-1}	E_b, kcal/mole	$\Delta_r H$, kcal/mole
27	$CH_3 + ^* \leftrightarrow CH_3^*$	1.00	0.0	1.00×10^{13}	38.0	-38.0
28	$CH_2 + ^* \leftrightarrow CH_2^*$	1.00	0.0	1.00×10^{13}	68.0	-68.0
29	$CH + ^* \leftrightarrow CH^*$	1.00	0.0	1.00×10^{13}	97.0	-97.0
30	$C + ^* \leftrightarrow C^*$	1.00	0.0	1.00×10^{13}	150.0	-149.0
31	$2CO^* \leftrightarrow C^* + CO_2^*$	2.40×10^{12}	31.0	4.17×10^9	0.0	31.9
32	$O^* + Z \leftrightarrow ^* + ZO$	0.2×10^3	1.4	0.2×10^3	1.4	0.0

The site densities were assumed: for the Pt - 1.65×10^{-5} mol/m^2, for ZO – 1.65×10^{-2}.

Generally, the simulation is shown to be in good agreement with the experimental results, yet for the first 20 seconds the predicted temperature in the front part of the monolith increases slightly faster than the experimentally measured temperature. During this time, the hottest part was found to be at the length of 0.01m in the experiment and at 0.005m in the simulations. The error bar of the measurements certainly existed, but the slower temperature rise in the experiments in comparison with the 2-D modeling data was also noted earlier by Schwiedernoch et al. [213]. The numerically predicted concentrations of components in the product gas (Figure 26a) are compared with the experimentally derived data (Figure 26b).

Again, a fair qualitative agreement between the measured and simulated species profiles is observed. During light-off, carbon dioxide, as a product of complete oxidation of methane, appears initially. Then, synthesis gas selectivity slowly increases with rising temperature. On the contrary to the data for Rh - loaded monolith [213], where CO formation starts before hydrogen formation, hydrogen is detected first during the ignition of Pt/Ce-Zr-La/α-Al$_2$O$_3$ monolith. And again, the simulated gas phase composition reaches steady state faster than in the real transient experiment. This fact was also observed by Schwiedernoch et al. [213]. A special study should be undertaken to clarify the matter discussed. However, the detailed reactor and reaction models can be applied in studying non-stationary/non-equilibrium area to obtain fine details of chemical processes. Detailed simulations allow insights into processes which could not be gained through experiments alone, particularly when extreme reaction conditions (such as very high temperatures) are involved.

Simulation of the thermal behavior of the ceramometal monolithic catalyst (Figure 27) shows the transient development of the gas temperature (left) and the solid temperature (right) profiles during the light-off process occurred on a time scale of seconds. A large difference between the gas and solid phase temperatures is observed. This agrees with results of IR thermography and thermocouple measurements performed by Basini et al. [215, 216] which revealed a large temperature difference between the surface and the gas phase, the surface temperature even exceeding the adiabatic gas temperatures.

Complex dynamic behavior of the surface species is revealed by the simulation data. In spite of the fact that the reactants' residence time in the operational conditions is of the order of milliseconds, the vacancies coverage reaches steady state rather slowly (Figure 28, left) due to the multi-step heterogeneous reaction mechanism. Calculated molar gas-phase concentrations at the steady state of methane oxidation in the monolith (Figure 28, right) show that the main chemical transformations occurred within 10% of length at the monolith entrance.

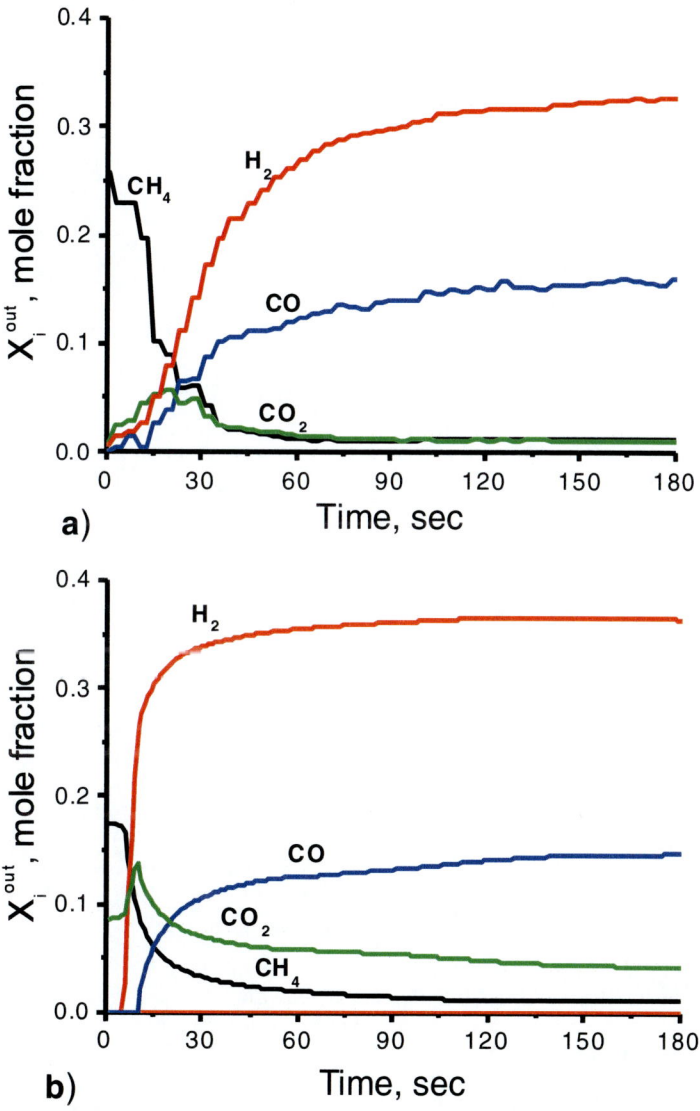

a)

b)

Figure 26. Time dependencies of main components in the product gas. Feed mixture: 26.7 %. CH_4 in air, T_o=400°C, P=1atm, u_o=0.34 m (STP)/s. a) - experimental data, b) - modeling results. Adapted with permission from Ref 211. Copyright 2007 Elsevier B.V.

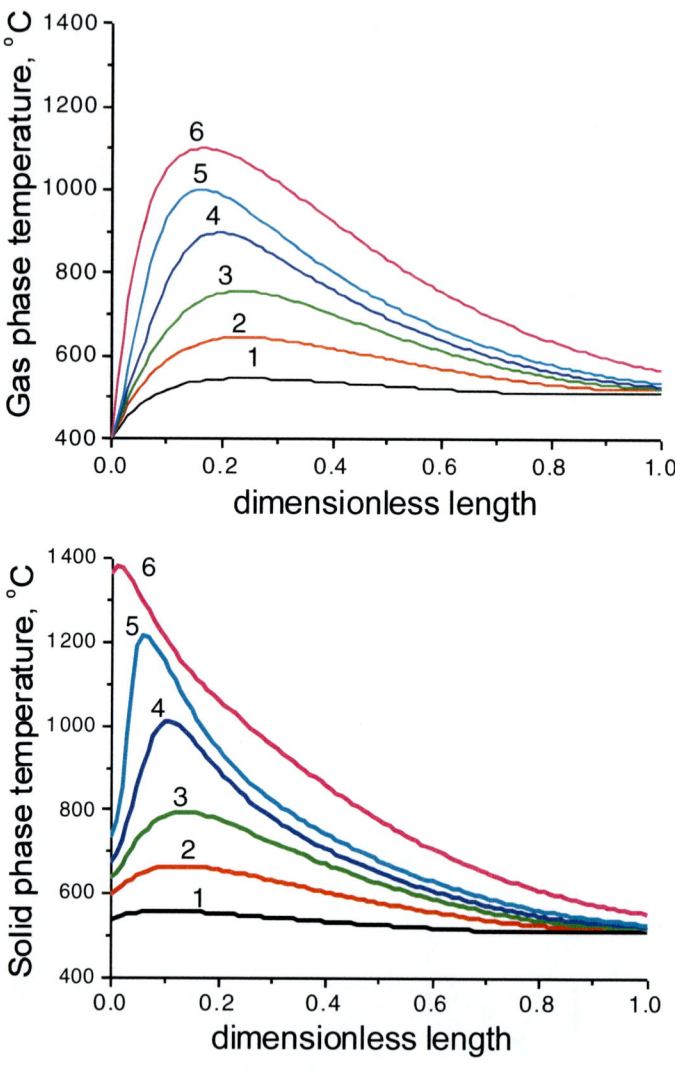

Figure 27. The temperature evolution in ceramometal monolith (30 PPI) during light-off: 1-in 0.8s, 2-2.4s, 3-4s, 5-5.6s, 5-7.2s, 6-12s. Feed mixture: 25 mol.%. CH_4 in air, $T_o=400^0C$, P=1atm, $u_o=0.5$ m (STP)/s.

Besides the fact that the high surface temperature is a major threat to the stability of supported noble metal catalyst, it is possible that the catalyst generates radicals involved in the chain reaction propagation on the surface, in the film at the gas-solid interface or even in the gas phase [23, 217].

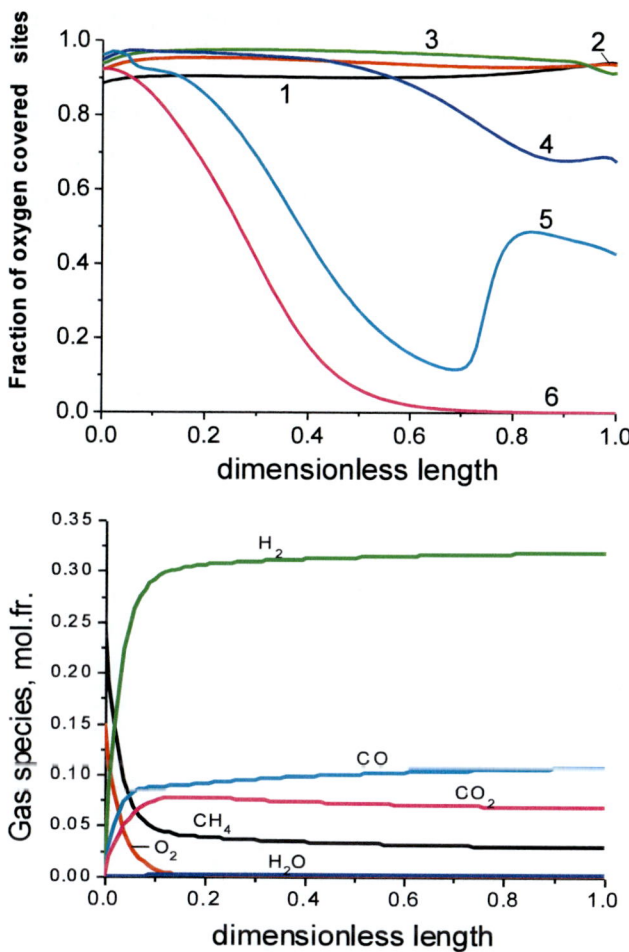

Figure 28. Fraction of oxygen-covered surface sites for the simulated light-off in ceramometal monolith (left): 1-in 0.8 s, 2-2.4 s, 3-4 s, 5-5.6 s, 5-7.2 s, 6-12 s, axial distribution of the gas phase species in steady state (right).

It is generally believed that at the atmospheric pressure homogeneous reaction paths play a minor role in the methane partial oxidation. To estimate the occurrence of gas phase chemistry under operational conditions typical for the methane partial oxidation, we applied a detail kinetic model (see Table 4) based on a free radical mechanism proposed by Berger and Marin [218]. The gas-phase mechanism contains 40 reversible elementary free-radical reactions with 13 molecules and 10 radicals. General features of the reaction mechanism do not depend on the total pressure. Effect of pressure was taken into account by the variation of kinetic parameters of unimolecular reactions. The curve of the pseudo-first order rate coefficient k_u for an unimolecular reaction $A + M \rightarrow A_1 + A_2 + M$ with $r = k_u C_A$, against pressure is called the fall-off curve of the unimolecular reaction. The value of a collision partner M for species A, known as the third-body concentration C_M ("bath gas"), consists of all the species in the reaction mixture. The third-body concentration C_M is commonly used in spite of the term "pressure" falloff [73]. The value for C_M was taken as the weighed sum of concentrations of all the molecular species. The weight factor w_i takes into account the relative collision efficiencies (assumed to be independent of temperature and reaction) of third bodies [218]:

$$C_M = \sum w_i C_i$$

The values of the weight factors required for the calculation of the concentration of third bodies were H_2O - 6.5, CH_4 - 6.5, CO_2 - 1.5, CO - 0.75, O_2 - 0.4, N_2 - 0.4, relative to hydrogen. Other weight factors were equal to unity [219].

The first order approximation is assumed for k_u:

$$k_u = \frac{kC_M}{1 + kC_M} k_\infty ,$$

where k_∞ is the rate coefficient for unimolecular reactions at a high-pressure limit. At sufficiently high pressure, that is, when $kC_M > 1$, k_u is independent of the total pressures, and approaches k_∞ - the rate coefficient for unimolecular reactions at a high-pressure limit. At low pressure, that is, when $kC_M < 1$, k_u becomes proportional to the total pressure, or the total concentration of "third bodies": $k_u = kk_\infty C_M = k_0 C_M$, with $k_0 = kk_\infty$

The Arrhenius parameters in Table 4 are represented at one of the limiting pressures, either k_∞ (at a high-pressure limit) or k_0 (at a low-pressure limit). For the pressure range of 0.1-2 MPa, the unimolecular reactions 16 and 36 are assumed to be completely in the low-pressure regime. In this case, the pseudo-first order rate coefficient for a unimolecular reaction would be $k_u = k_0 C_M$. The reactions 30, 31, and 33 are assumed to be in the high pressure regime and are therefore represented in Table 4 without the third body M. The decomposition of $CH_3O^\bullet$ (reaction 11) and the decomposition of H_2O_2 (reaction 38) are also mainly in the low-pressure regime, but these are influenced by pressure falloff to some extent at higher pressures:

$$k_u = k_0 \frac{C_M}{1 + C_M \alpha_1 \exp(\alpha_2/(RT))}.$$

For reactions 1, 10, 22, 27 the falloff data were accurately fitted by correlations:

$$k_u = k_\infty \frac{C_M^{\alpha_4} \exp(\alpha_1 + \alpha_2 T + \alpha_3 T^2)}{1 + C_M^{\alpha_4} \exp(\alpha_1 + \alpha_2 T + \alpha_3 T^2)}.$$

The corresponding falloff coefficients α_i are listed in Table 5.

The temperature along the monolith and time dependencies of main components at the exit calculated with this model are shown in Figure 29. These results imply that for the typical partial oxidation of methane-air mixture and operational temperatures, the local gas phase reactions may occur in the presence of oxygen in the gas phase. The calculated concentrations of gas-phase species indicate that the interplay between homogeneous gas phase and catalytic surface reactions may affect considerably the spatial distribution of both the temperature and concentration of species at the monolith entrance. High temperatures help to overcome rather high activation barriers of homogeneous gas-phase reactions involving methane molecules, making these reactions "competitive" with catalytic pathways. For the liquid fuels which are much more reactive, homogeneous side reactions can change considerably the process selectivities and temperature, and be even responsible for flames and explosions at high temperature conditions.

Methane is the most favorable fuel to be used in studies of the reaction behavior based on the detailed reaction kinetics. Information about kinetics of the partial oxidation or steam reforming reactions of higher hydrocarbons and oxygenates is presently only scarcely available. Further studies of coupling between catalytic and non-catalytic reactions using a combination of experimental investigations and mathematical simulations are very important from scientific interest and practical points of view.

Table 4. Kinetic model for the gas-phase partial oxidation of methane to synthesis gas in the absence of a catalyst

№	Reactions	k_0 ^{a}	E_a, kJ/mol	A/RT
1	$CH_4 + M \leftrightarrow CH_3^{\bullet} + H^{\bullet} + M$	0.24×10^{17}	438.98	-6.3
2	$CH_4 + O_2 \leftrightarrow CH_3^{\bullet} + H_2O^{\bullet}$	0.398×10^{8}	223.03	0.1
3	$CH_4 + H^{\bullet} \leftrightarrow CH_3^{\bullet} + H_2$	0.473×10^{8}	50.31	2.0
4	$CH_4 + O^{\bullet} \leftrightarrow CH_3^{\bullet} + OH^{\bullet}$	0.173×10^{9}	49.12	12.0
5	$CH_4 + OH^{\bullet} \leftrightarrow CH_3^{\bullet} + H_2O$	0.659×10^{8}	34.54	4.8
6	$CH_4 + HO_2^{\bullet} \leftrightarrow CH_3^{\bullet} + H_2O_2$	0.128×10^{8}	88.18	5.7
7	$CH_3^{\bullet} + O_2 \leftrightarrow CH_2O + OH^{\bullet}$	0.396×10^{6}	54.29	37.8
8	$CH_3^{\bullet} + O_2 \leftrightarrow CH_3O^{\bullet} + O^{\bullet}$	0.102×10^{10}	151.30	11.7
9	$CH_3^{\bullet} + HO_2^{\bullet} \leftrightarrow CH_3O^{\bullet} + OH^{\bullet}$	0.255×10^{8}	0.00	23.6
10	$2CH_3^{\bullet} + M \leftrightarrow C_2H_6 + M$	0.329×10^{7}	-11.34	6.9
11	$CH_3O^{\bullet} + M \leftrightarrow CH_2O + H^{\bullet} + M$	0.383×10^{9}	81.12	7.8
12	$CH_2O + O_2 \leftrightarrow HO_2^{\bullet} + CHO^{\bullet}$	0.282×10^{9}	184.27	8.7
13	$CH_2O + OH^{\bullet} \leftrightarrow CHO^{\bullet} + H_2O$	0.951×10^{9}	7.74	13.5
14	$CH_2O + HO_2^{\bullet} \leftrightarrow CHO^{\bullet} + H_2O_2$	0.461×10^{7}	43.62	14.3
15	$CH_2O + CH_3^{\bullet} \leftrightarrow CHO^{\bullet} + CH_4$	0.266×10^{7}	13.39	8.6
16	$CHO^{\bullet} + M \leftrightarrow CO + H^{\bullet} + M$	0.835×10^{8}	47.07	3.8
17	$CHO^{\bullet} + O_2 \leftrightarrow CO + HO_2^{\bullet}$	0.305×10^{8}	13.74	10.2
18	$CO + HO_2^{\bullet} \leftrightarrow CO_2 + OH^{\bullet}$	0.474×10^{8}	73.95	33.5
19	$C_2H_6 + H^{\bullet} \leftrightarrow C_2H_5^{\bullet} + H_2$	0.223×10^{9}	44.10	4.5
20	$C_2H_6 + OH^{\bullet} \leftrightarrow C_2H_5^{\bullet} + H_2O$	0.230×10^{9}	18.60	7.4

№	Reactions	k_0 [a]	E_a, kJ/mol	A/RT
21	$C_2H_6 + CH_3^{\bullet} \leftrightarrow C_2H_5^{\bullet} + CH_4$	0.874×10^9	97.64	2.5
22	$C_2H_5^{\bullet} + M \leftrightarrow C_2H_4 + H^{\bullet} + M$	0.317×10^{15}	195.98	0.8
23	$C_2H_5^{\bullet} + O_2 \leftrightarrow C_2H_4 + HO_2^{\bullet}$	0.377×10^5	-1.56	7.2
24	$C_2H_4 + H^{\bullet} \leftrightarrow C_2H_3^{\bullet} + H_2$	0.542×10^9	62.36	4.9
25	$C_2H_4 + OH^{\bullet} \leftrightarrow C_2H_3^{\bullet} + H_2O$	0.205×10^8	24.86	7.8
26	$C_2H_4 + CH_3^{\bullet} \leftrightarrow C_2H_3^{\bullet} + CH_4$	0.416×10^7	46.56	2.9
27	$C_2H_3^{\bullet} + M \leftrightarrow C_2H_2 + H^{\bullet} + M$	0.200×10^{15}	166.28	0.1
28	$C_2H_3^{\bullet} + O_2 \leftrightarrow C_2H_2 + HO_2^{\bullet}$	0.121×10^6	0.00	6.5
29	$C_2H_3^{\bullet} + O_2 \leftrightarrow CH_2O + CHO^{\bullet}$	0.542×10^7	0.00	44.8
30	$C_2H_5^{\bullet} + CH_3^{\bullet} \leftrightarrow C_3H_8$	0.105×10^9	0.00	3.8
31	$C_2H_4 + CH_3^{\bullet} \leftrightarrow C_3H_7^{\bullet}$	0.109×10^7	35.64	-0.4
32	$C_3H_8 + H^{\bullet} \leftrightarrow C_3H_7^{\bullet} + H_2$	0.238×10^{10}	30.44	4.9
33	$C_3H_7^{\bullet} \leftrightarrow C_3H_6 + H^{\bullet}$	0.433×10^{15}	157.69	1.0
34	$C_3H_7^{\bullet} + O_2 \leftrightarrow C_3H_6 + HO_2^{\bullet}$	0.119×10^7	11.01	7.3
35	$O_2 + H^{\bullet} \leftrightarrow OH^{\bullet} + O^{\bullet}$	0.728×10^9	77.91	14.3
36	$O_2 + H^{\bullet} + M \leftrightarrow HO_2^{\bullet} + M$	0.150×10^3	-6.98	6.4
37	$HO_2^{\bullet} + HO_2^{\bullet} \leftrightarrow O_2 + H_2O_2$	0.121×10^7	14.92	5.6
38	$H_2O_2 + M \leftrightarrow OH^{\bullet} + OH^{\bullet} + M$	0.967×10^{10}	159.66	14.2
39	$OH^{\bullet} + H_2 \leftrightarrow H_2O + H^{\bullet}$	0.304×10^8	29.08	2.9
40	$HO_2^{\bullet} + H^{\bullet} \leftrightarrow OH^{\bullet} + OH^{\bullet}$	0.506×10^9	3.66	26.2

[a] k_0 is expressed in s^{-1} or $m^3 mol^{-1} s^{-1}$ or $m^6 mol^{-2} s^{-1}$. The relation between the affinity A, the forward $(\vec{r})$ and backward reaction rate $(\bar{r})$ is $\ln(\vec{r}/\bar{r}) = A/RT$. Conventional two-parametric Arrhenius dependency was used to express the rate constants.

Table 5. Falloff coefficients

No	α_1	α_2	α_3	α_4
1	-0.4904	-2.3380E-03	1.9720E-07	0.75
10	3.9690	-4.8740E-03	3.8020E-07	0.55
22	1.8920	-5.0580E-03	7.2850E-07	0.7
27	0.4197	-2.7810E-03	3.5420E-07	0.75
11	2.3940E-06	2872.3		
38	6.0030E-05	2800.1		

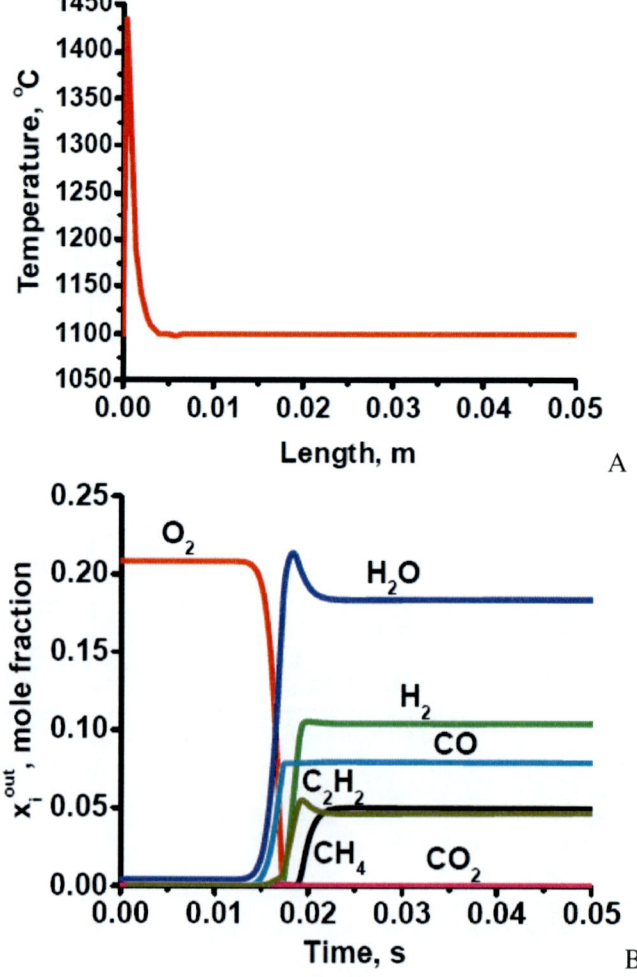

A

B

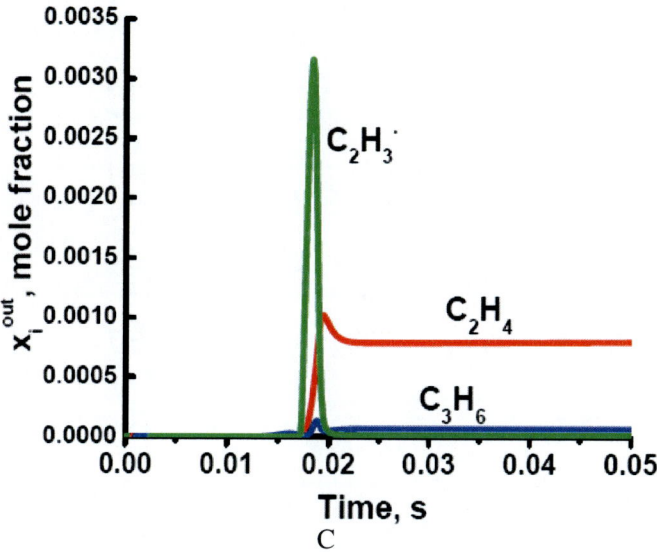

Figure 29. The simulated gas temperature along a monolith channel in steady state (A) and time dependencies of the main gas species at the exit (B and C). Feed mixture: 25 %. CH_4 in air, P=1atm, u_o=0.7 m/s, T_o=1100°C, d_{eq}=1.15 mm.

Chapter 5

SOME APPLICATION OF STRUCTURED CATALYSTS FOR SYNGAS GENERATION FROM HYDROCARBONS AND OXYGENATES IN THE PILOT SCALE REACTORS

The precise experimental details for the structured catalysts used in a range of applications have been described elsewhere [83, 116-124, 188-192]. We present here only the main results and ideas of a general interest and significance.

5.1. SYNGAS GENERATION IN A MONOLITHIC REACTOR

5.1.1. Composite Ceramometal Monolith

Oxidative reforming of hydrocarbon fuels. For isooctane – air mixture, the ignition temperature of this catalyst performance is 275°C [189], while the temperature ~ 410°C was required to light- off the reaction of methane partial oxidation. For a feed containing 24% of methane in air and at 0.084 s (STP) contact time, composite ceramometal monolith provides a higher methane conversion combined with a high CO and H_2 yield compared to the honeycomb catalyst on corundum substrate (96, 95, 94 % against 84, 91, 87%, respectively). This is due to the optimal textural properties of the composite cermet monolith characterized by the high mass transfer rates to the catalytic sites.

It is known that a catalyst lifetime is limited by thermal stresses and deactivation. Catalyst durability can be improved by alleviating the axial temperature gradients which result in the thermal stresses [220]. The temperature profile in an adiabatically operated monolith is characteristic of the reforming chemistry, catalytic composition, and properties of the monolithic substrate. The cermet monolithic substrate has a higher heat conductive transfer compared to that for corundum substrate. A difference of about 200°C between the front and back faces of the cetmet monolith was detected in the steady state when testing both isooctane (Figure 30, left) and methane [117, 118, 189], in spite of the fact that the reaction heat evolved in the partial oxidation of isooctane to syngas is higher then that in the methane conversion.

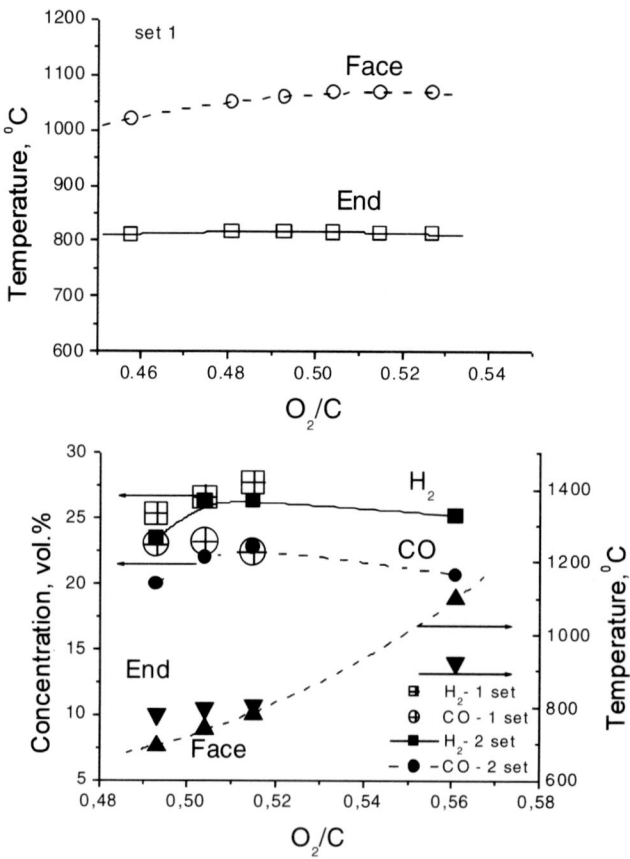

Figure 30. Isooctane (0.350 kg/h) partial oxidation performed on the composite 7.8 %wt LaNiPt/ 0.87 %wt LaRu/cermet monolith. Figure on the left and open symbols on the right exhibit experimental data in case of 0.02 % wt. Rh addition on the inlet part of the catalytic monolith (set 1).

The rear-face temperatures responded just a little to variation of O_2/C ratio in the feed. The temperature profile is also affected by variation of the active component composition along the monolith [221, 222]. Addition of 0.02 wt% Rh into the inlet part of the cermet monolith with basic active component comprised of 7.8 %wt. LaNiPt/0.87 %wt. LaRu provides a higher exothermicity of the local reactions in the presence of gas-phase oxygen. However, the product gas composition is determined by thermodynamic equilibrium at the exit temperatures, which are nearly the same for considered catalysts. C_2 -C_3 components were not detected in the product gas that can be explained by a minor role of homogeneous reactions in the cermet monolith geometry. Besides, a high thermal stability and mechanical strength make composite ceramometal monoliths very attractive for the practical application.

5.1.2. Metallic Monolith

For this type of monoliths, both partial oxidation and authothermal reforming processes with methane, isooctane and gasoline have been studied experimentally [122, 123, 189-192]. The experimental data on the oxidative reforming of liquid fuels in a pilot scale reactor are shown in Tables 6, 7 and Figures 31-33.

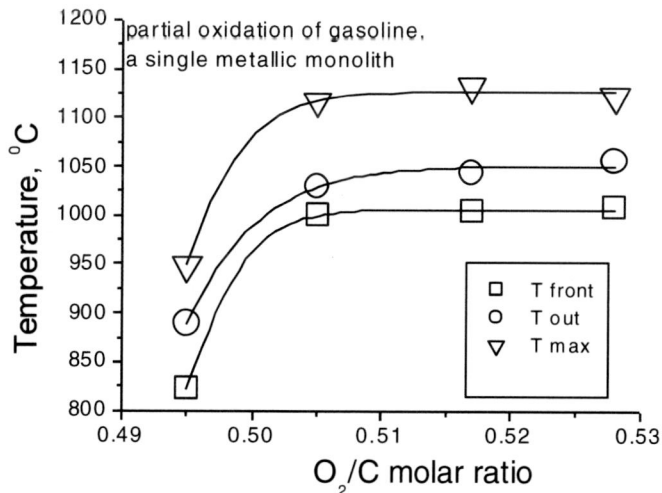

Figure 31. Effect of O_2/C ratio on the catalyst temperature; gasoline flow rate 0.893 kg/h.

Table 6. Experimental data on the oxidative reforming of liquid fuels over metallic monolith

Flow rates	Isooctane: 0.757 kg/h Air: 3.00 m^3(n)/h Q=3.15 m^3(n)/h O$_2$/C=0.53 Tin = 205 ^{o}C	Isooctane: 0.757 kg/h Air: 3.00 m^3(n)/h Water: 0.735 kg/h Q=4,1 m^3(n)/h O$_2$/C=0.53 H$_2$O/C=0.8 Tin= 265 ^{o}C	Gasoline: 0.893 kg/h Air: 3.31 m^3(n)/h Q=3.51 m^3(n)/h O$_2$/C=0.48 Tin = 220 ^{o}C	Gasoline: 1.86 kg/h Air: 6.90 m^3(n)/h Q=7.32 m^3(n)/h O$_2$/C=0.48 Tin = 206 ^{o}C
Contact time, s	0.105	0.081	0.19	0.09
Superficial velocity, m/s (STP)	0.38	0.5	0.42	0.9
Front-face catalyst temperature, ^{0}C	1033-1067	985-957	1081	1117
Back-face catalyst temperature, ^{0}C	903-951	913-892	963	1000
Synthesis gas (dry) composition: % mol				
H$_2$	26-27.3	31-31.6	23.9	24.8-24.7
CO	22.2-24.8	19.2-17.7	28.6	27.7-28.7
CO$_2$	0.4-2.7	4.5 -6.0	trace	trace -0.4
N$_2$	49-53.0	44.5-49.1	50.3	51.6-51.7
CH$_4$	trace – 0.2	0.1 - 0.3	0.4	trace -0.1
C$_2$ - C$_3$	0	0	0	0

Table 7. Data on autothermal reforming of gasoline over two metallic monoliths

Parameter	Run 1 (O$_2$/C=0.48; H$_2$O/C=1.17)			Run 2 (O$_2$/C=0.52; H$_2$O/C=0.89)			Run 3 (O$_2$/C=0.48; H$_2$O/C=0.86)		
	Experiment	Predicted data		Experiment	Predicted data		Experiment	Predicted data	
		Isothermal reactor at T$_{out}$	Adiabatic reactor at T$_{in}$		Isothermal reactor at T$_{out}$	Adiabatic reactor at T$_{in}$		Isothermal reactor at T$_{out}$	Adiabatic reactor at T$_{in}$
Gasoline, kg/h	1.86			0.893			1.86		
Air, m^3/h (STD)	6.90			3.58			6.98		
Water, kg/h	3.13			1.14			2.3		
Q, m^3/h (STD)	11.22			5.17			10.18		
Contact time, (at STD) s	0.059			0.128			0.065		
Superficial velocity, m/s (STD)	1.36			0.63			1.23		
Feed temperature (T$_{in}$), °C	260			280			250		
Front (Back)-face temperature (T$_{out}$). °C	763-860			1050-1000			958-898		
Synthesis gas (dry) composition: % mol									
H$_2$	31.7	31.8	32.0	27.5	27.9	28.1	30.3	30.4	30.6
CO	15.6	15.9	15.6	17.4	18.1	17.6	18.2	16.0	15.7
CO$_2$	9.6	8.4	8.6	7.1	6.4	6.6	6.3	6.5	6.8
N$_2$	45.1	43.4	43.3	49.4	47.3	47.0	46.3	44.3	44.2
		0.0036	0.0057	trace	0.0002	0.0004	trace	0.0025	0.0047

Table 7. (Continued)

Parameter	Run 1 (O_2/C=0.48; H_2O/C= 1.17)			Run 2 (O_2/C=0.52; H_2O/C=0.89)			Run 3 (O_2/C=0.48; H_2O/C= 0.86)		
	Experiment	Predicted data		Experiment	Predicted data		Experiment	Predicted data	
		Isothermal reactor at T_{out}	Adiabatic reactor at T_{in}		Isothermal reactor at T_{out}	Adiabatic reactor at T_{in}		Isothermal reactor at T_{out}	Adiabatic reactor at T_{in}
CH_4 $C_2 - C_3$	trace 0			0			0		
Predicted adiabatic temperature (back face), °C			840			977			869

Radial and axial temperature gradients developed in monoliths may lead to corresponding variations in both catalyst efficiency and selectivity, which are known to be very sensitive to temperature. It is explained by the fact that an increase in temperature favors the intrinsic kinetic constant with respect to diffusional constant [223]. At the fixed fuel flow rate, increasing O_2/C molar ratio in the feed enhances exothermic reactions in the reforming process. Heat conduction in the metallic supports of structured catalysts can provide an effective alternative mechanism to remove the heat generated by strongly exothermic reactions and transfer it into zone where endothermic reactions occur, thereby providing a smoothing effect on the catalyst temperature profile (Figure 31).

The effect of gasoline feed rate on the metallic monolith performance in the partial oxidation and autothermal reforming processes is demonstrated by the results listed in Table 6. The oxygen-to-carbon ratio in the feed was slightly lower than required by syngas stoichiometry. The molar ratio of hydrogen to carbon monoxide (H_2/CO) in syngas produced in the partial oxidation of gasoline was about 0.84 –0.9 at the theoretical value 0.93. In the experiments with differing throughputs, the feed composition was kept at nearly constant level. The composition of converted stream was independent of the throughput. A small difference in the catalyst temperatures was also observed, in spite of different heat generated in the reaction, which is dependent on the feed velocity.

In a good approximation, the metallic monoliths operate adiabatically over a range of O_2/C ratios in the gasoline-air mixture (Figure 32). The equilibrium curves for the species concentrations and adiabatic temperature rise at the exit temperature and constant pressure (101.325 kPa) have been determined by using calculations of thermodynamic equilibrium with a model gasoline fuel. Agreement of calculations with the experimental data is rather good, especially considering a complexity of the system being modeled and the assumptions made.

The experimental data along with results of thermodynamic calculations for the reforming process with steam addition to the reaction mixture are presented in Table 7. The calculations of equilibrium at the back face temperature (isothermal reactor) suggest also that the product gas composition in the experiments is determined by the outlet temperature. This is valid even in the case when the back face temperature exceeds that of the front face one (see run 1 in Table 7) at a high throughput. Hence, agreement between the experimental data for the products distribution and thermodynamic predictions is good in this case as well.

It should be noted that the fuel-oxygen ratio is the most important tunable parameter for the reformation of the high energy density gasoline fuel. The variation in the air flow rate for only two percents at the same feed rate of gasoline causes a pronounced impact on both the adiabatic temperature and reformate composition.

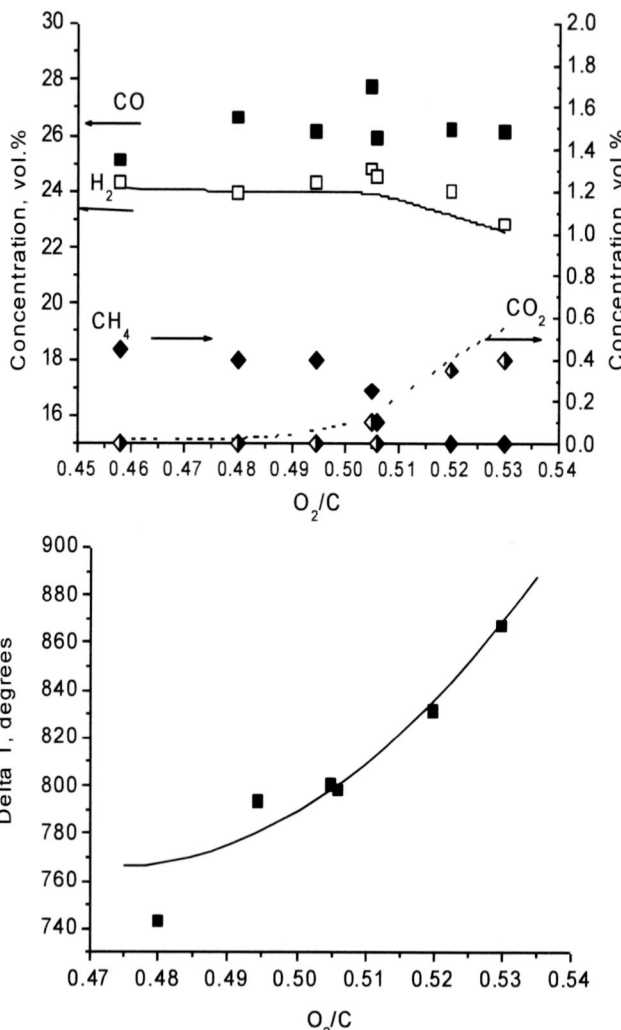

Figure 32. Gasoline partial oxidation by air over two metallic monoliths (1 %wt LaNiPt/ 1 %wt LaRu/ LaCeZrO$_x$ washcoat) Experimentally measured values (symbols) and thermodynamic predictions (curves) for the product gas (left) and adiabatic temperature rise (right) in the partial oxidation of gasoline versus O$_2$/C molar ratio in the feed stream. The feed temperature 190-210°C.

According to the experimental observations, enhancing exothermicity of the gasoline partial oxidation reaction by increasing O_2/C molar ratio in the feed causes an increase of the outlet temperature while the front-face temperature changes only slightly (Figure 33).

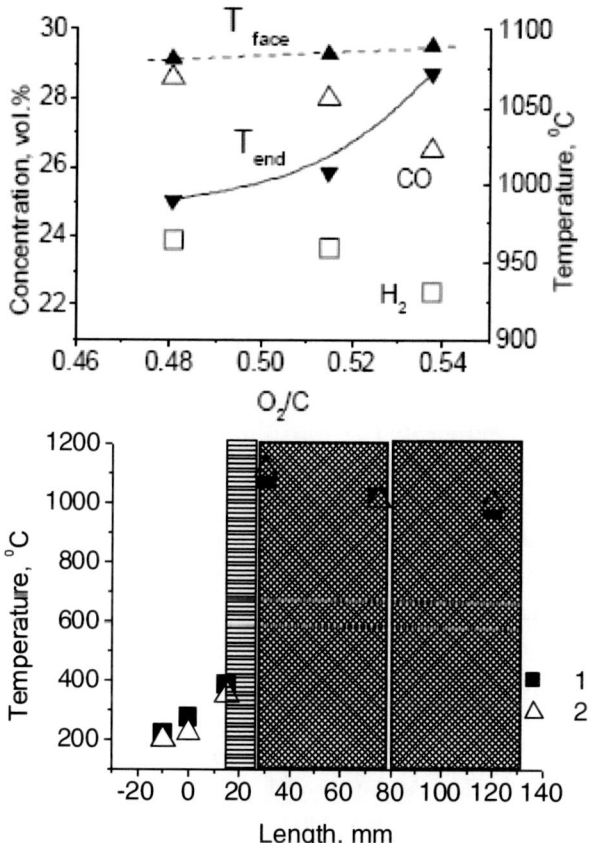

Left: Effect of air-fuel ratio (expressed by O_2/C mole ratio in the feed) on hydrogen and CO concentration (open symbols) in syngas and the temperature (filled symbols). Gasoline flow rate – 0.893 kg/h.

Right: The temperature profiles along the reactor axis at the steady state and O_2/C =0.48.

Mode 1 - 0.893 kg/h gasoline, 3.31 m³/h air (STD) (O_2/C =0.48), U_o = 0.42 m/s and 0.19 s contact time, T_{in}=220°C; mode 2 - 1.86 kg/h gasoline, 6.90 m³/h, U_o = 0.9 m/s and 0.09 s contact time, T_{in}=206°C.

Figure 33. Gasoline partial oxidation by air over two metallic monoliths.

Comparing to the ceramic monolith, the thermal axial (longitudinal) gradient is minimized in the metallic monoliths. A similar trend was observed for the radial (transversal) catalyst temperature. A maximum difference of 60 degree was detected between the axial and wall temperatures in the reactor with the metallic monolith against the difference of 90°C for microchannel cermet catalyst. Therefore, the catalyst bed with a high thermal conductivity may act as a heat exchanger through the proper axial distribution of the process heat. For the bed comprised of two stacked metallic monoliths, at a constant oxygen-to-carbon ratio, the flow rate practically does not affect the axial temperature profile (Figure 33). Hence, metallic monolithic catalysts developed at Boreskov Institute of Catalysis ensure an extremely efficient transfer of reaction heat within the reactor due to a proper choice of support material and monolith geometry.

5.1.3. Ignition Studies

The catalytic processes, which can be run nearly autothermally and adiabatically, exhibit extremely fast variations in the temperature during ignition period causing severe thermal shock to the monolith. A detailed knowledge of ignition is of prime importance due to both economic and safety concerns. The ignition performance may differ from one catalyst to another. Thus, it is necessary to have a thorough study on individual catalysts. Here the results on ignition characteristics of monolithic catalysts comprised of 3-7 wt.% $LaNi_{0.9}Pt_{0.1}O_x$/5-10 wt.% $Zr_{0.8}Ce_{0.2}O_2$ active component supported on full-sized corundum (Figure 12) and fechraloy foil (Figure 9) substrates are presented.

Figure 34 shows schematic representation of an assembly of the catalytic monolith (hexagonal corundum or fechraloy foil –based catalytic monolith) and front thermal shield loaded into experimental tubular stainless-steel reactor. The axial temperatures at selected points along the monolith were scanned by means of thermocouples located in the central channel plugged with alumina-silica fibers. In the test experiments, the feed (27 vol.% of natural gas in air) flow rate was varied in the range corresponding to contact times of 0.02-0.2 s (STP). Gas composition was determined by gas chromatography (GC) and quadrupole mass spectroscopy (QMS) in the transient mode. The short contact time reactor was initially heated up with hot air until the catalyst reached the temperature of 300-400°C, and then natural gas was introduced into the air stream. Both main components in the product gas and evolution of the catalyst temperatures at different points along the monolith length during reaction ignition over the corundum-based monolith are presented in Figure 35.

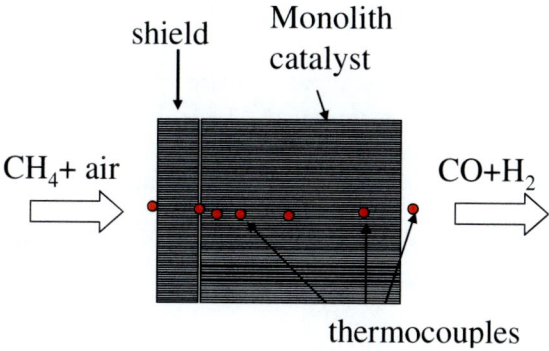

Figure 34. Schematic representation of the monoliths assembly in the experimental reactor.

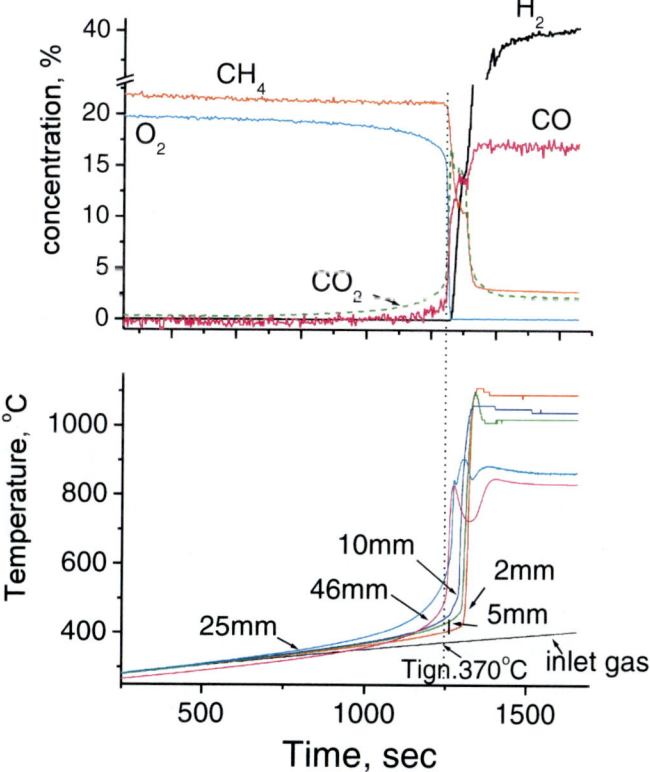

Figure 35. Main components in the product gas and evolution of the catalyst temperatures at different points along the monolith length during ignition of the partial oxidation reaction over the LaNiPt-CeZrO$_2$/corundum-based monolith. See text for details.

It can be seen that O_2 consumption and CO_2 formation starts already at 300°C before the ignition temperature of 370°C due to CH_4 combustion on the surface sites. At the ignition point, the rapid increase of the catalyst temperature and methane and oxygen consumption with syngas formation is observed. Initial temperature of the catalyst considerably affects the ignition features. Figure 36 shows the results of the experiments when natural gas was introduced into the hot air stream at temperatures equal to the catalyst preheat (initial) temperature.

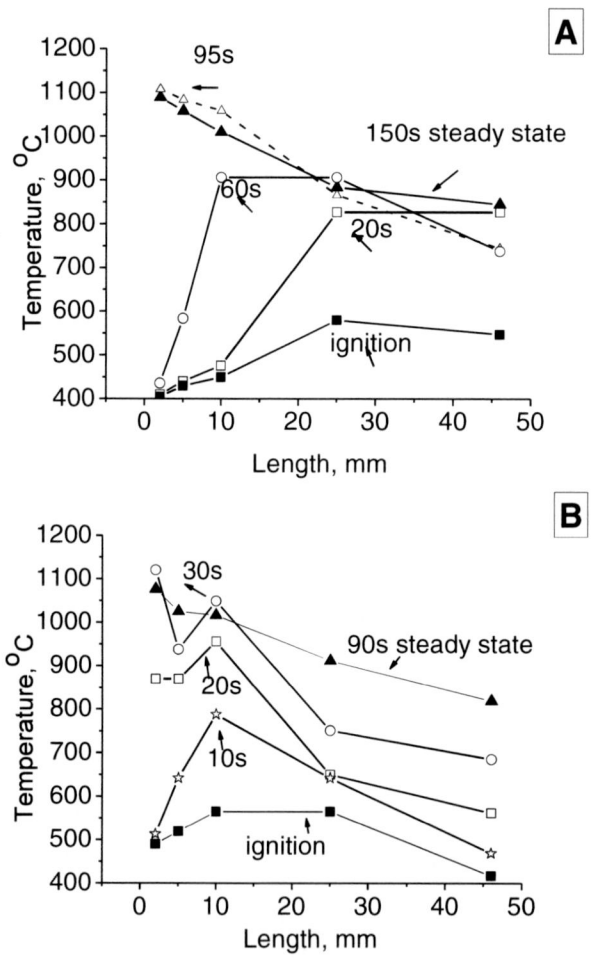

Figure 36. Evolution of the axial temperature profiles in the corundum-based monolith during start-up. Contact time - 0.1 s. Initial temperature: (A) - 300°C; (B) - 400°C.

At the initial temperature of 300°C, ignition starts in the rear part of monolith (Figure 36A). Then the front of maximal temperature moves to the entrance and at 90 s the temperature profile is stabilized. At higher temperature of both the catalyst and the feed, ignition starts in the central zone of the corundum-based monolith (Figure 36B). In this case, the ignition occurs more rapidly, and the front of maximal temperatures reaches the monolith entrance in 30 s. At higher flow rate (shorter contact time 0.02 s), the ignition starts in the rear part of the catalyst bed, then the temperature front moves quickly from the rear part to the inlet of the catalyst, and its maximum reaches a stationary position at 10 mm length from the entrance (Figure 37). The temperature profile stabilizes in 30 s.

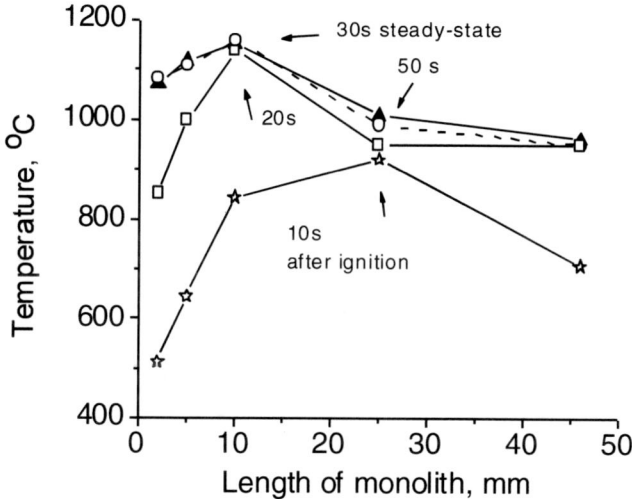

Figure 37. The axial temperature profiles in the selected points along the corundum-based monolith during start- up period. Contact time 0.02 s. Both the initial catalyst temperature and the feed temperature are equal to 400°C.

Effect of the thermal conductivity of the monolith substrate may be evaluated by comparison with the experimental data obtained for the metal-based monolith (see Figure 38.). For both catalysts, ignition starts in the rear part. However, the catalyst temperature reaches steady state more rapidly for the metallic substrate due to its high thermal conductivity. The temperature profiles at steady state are shown in Figure 39 for the catalytic monoliths based on corundum and metallic substrates for the flow rates corresponding to the contact times of 0.03 and 0.2 s. One can see that both a high temperature zone and a low temperature zone corresponding to the exothermic oxidation and endothermic reforming reaction routes are clearly distinguished at the contact time of 0.2 s. A steeper temperature profile is observed for the corundum-based monolith. High flow rate corresponding to the residence time of 0.03 s (STP) minimizes the observed differences caused mainly by thermal conductivities of substrates, thus making the steady state temperature profiles very similar for both corundum and fechraloy foil based monoliths.

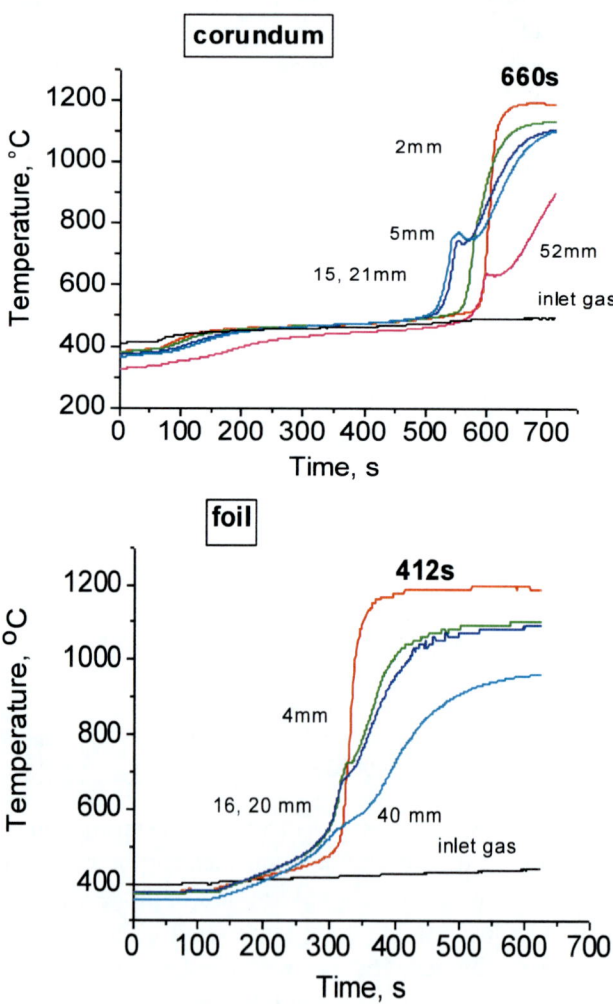

Figure 38. Time dependence of the temperature in the selected points along the monolith based on corundum (A) and fechraloy foil (B) during the ignition period. Contact time of 0.2 s. Both the initial catalyst temperature and the feed temperature are equal to 400°C.

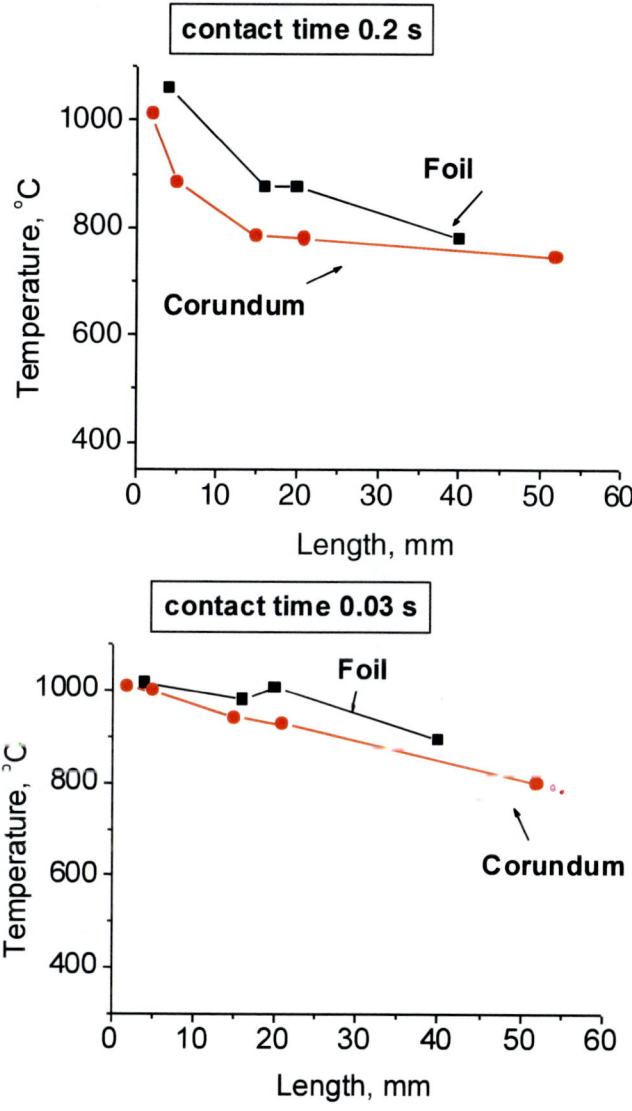

Figure 39. Effect of the material substrate and contact time on the temperature profile at the steady state of the methane partial oxidation reaction.

Effect of the residence time in the corundum-based monolith (tuned by the feed flow rate variation) on both methane conversion and syngas yield is shown in Figure 40. It can be seen that the feed flow rate may be varied within certain ranges without causing a considerable deterioration of the syngas generation process. Generally, chemistry, mass and heat transfer in the monolith affect the catalyst temperature, conversion and selectivity in the short contact time reactor. A higher heat generation in the oxidation zone occurs at increasing flow rate in the corundum-based monolith, which accompanied by a decrease in the contact time (Figure 40). This phenomenon is responsible for the higher catalyst temperatures observed (see Figure 40) at decreasing the residence time.

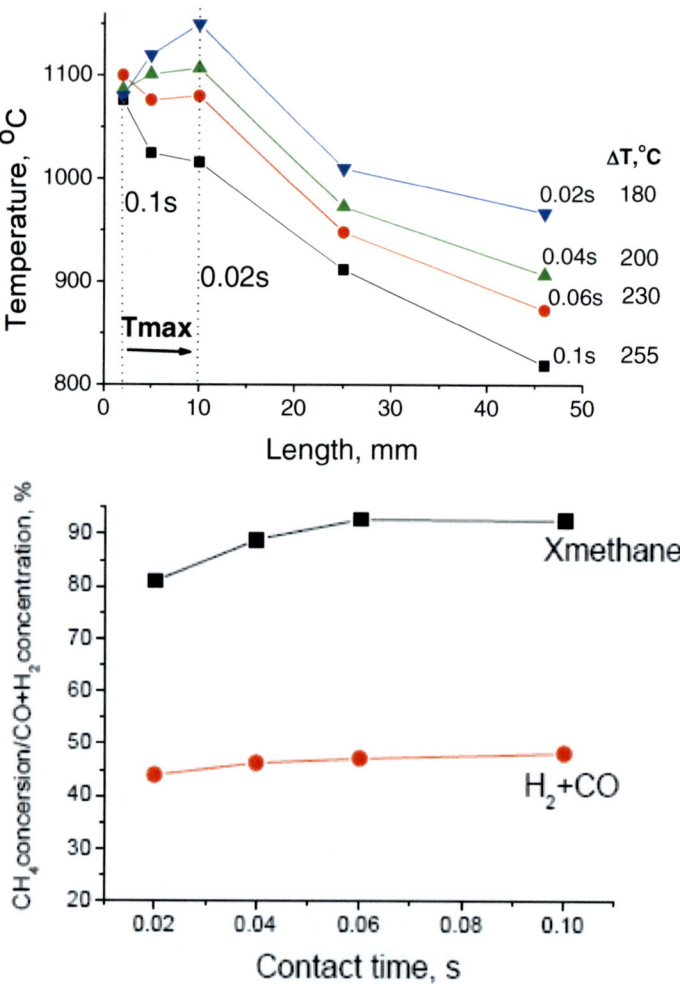

Figure 40. Effect of the residence time in the corundum-based monolith (caused by the feed flow rate variation) on the temperature profiles (left), methane conversion and syngas yield (right).

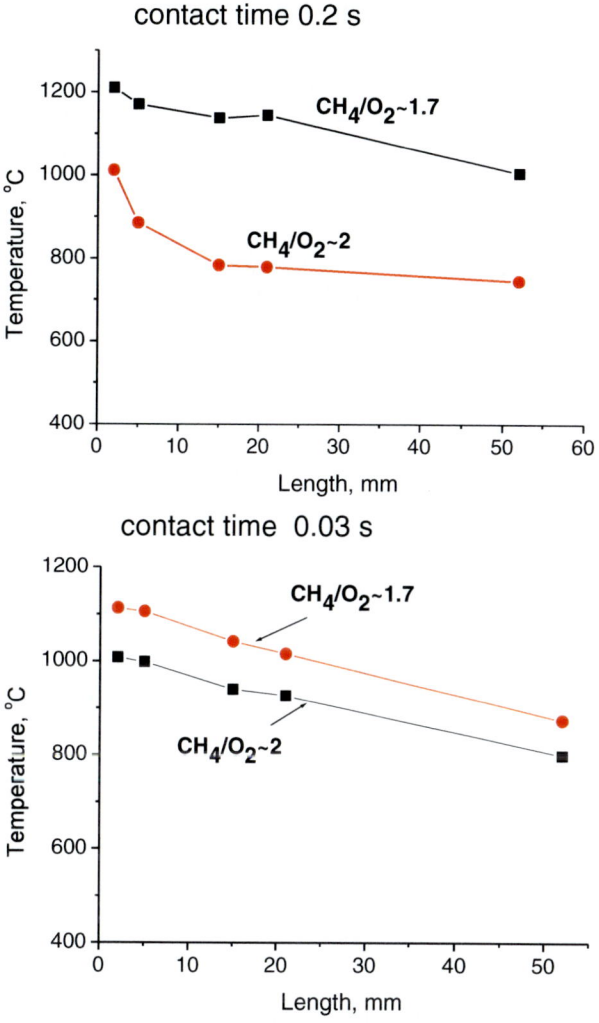

Figure 41. Influence of CH_4/O_2 ratio on the temperature of the corundum supported catalyst.

Location of the maximal temperature zone is also affected by the flow rate. At a slow flow rate (longer contact time), the exothermic reactions occur at the very entrance of the monolith. At the high flow rates, a zone of the maximal temperature shifts to the rear part of the monolith due to cooling the inlet part by incoming gas. The temperature gradient along the monolith (the difference between maximum and outlet temperatures, ΔT in Figure 40) tends to decrease with increasing the flow rate. This may be explained by dissipation of heat generated in the oxidation zone due to changing the interphase transport properties. Besides, the overall reaction mechanism could be changed also due to changing in the adsorption of reactants and co-reactants with temperature, including also a phenomenon broadly known as adsorption enhanced surface reconstruction [226].

The methane/oxygen ratio also affects the temperature profiles considerably. A high temperature and a low temperature zones corresponding to the exothermic oxidation and endothermic reforming reaction routes are clearly distinguished at the contact time of 0.2 s only, when oxygen concentration corresponds to the stoichiometric value (Figure 41). Excess oxygen results in a higher catalyst temperature, while smooth profiles are observed for both contact times of 0.2 and 0.03 s. As the result, both CH_4 conversion and syngas content in the product gas depend only slightly upon the air/natural gas ratio (Figure 42).

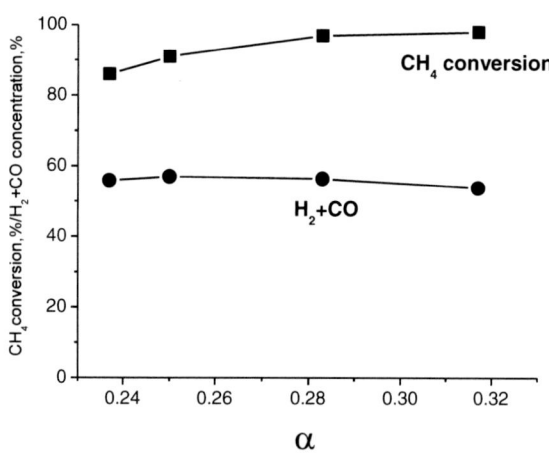

Figure 42. Dependence of methane conversion and syngas concentration on the ratio natural gas/air (α) at contact time 0.1 s on LaNiPt-CeZrO$_2$/corundum-based monolith (length 40 mm), feed preheat temperature 430 °C.

5.1.4. Effect of Pressure on Partial Oxidation of Methane

In some applications (gas turbines etc), partial oxidation of methane into syngas is perfomed at enhanced pressures. Study of the effect of pressure on the catalytic process is of a great impotance. As follows from Figure 43, the temperature of the monolith front edge increases with pressure weakly depending upon the feed rate. From the materials science point of view, front edge temperatures in excess of 1300 °C at high pressures are rather tough and are to dealt with by tuning the active component distribution along the monolith length. It was found that LaNiO$_3$ being used in the front layer of two stacked monoliths is much more stable to evaporation (Figures 44-47).

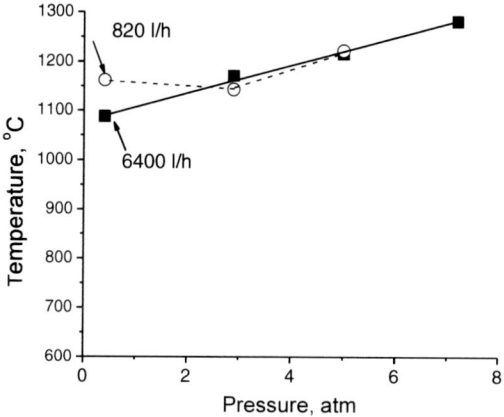

Figure 43. Dependence of the front end monolith temperature on the pressure and the feed rate. LaNiPt-CeZrO$_2$/corundum-based monolith (length 40 mm), feed preheat temperature 430 °C.

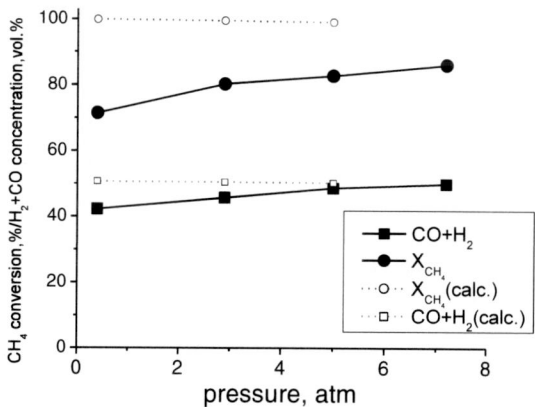

Figure 44. Pressure dependence of methane conversion and syngas concentration. The feed rate 6400 l/h (25 % CH$_4$ in air). Catalyst package is comprised of two stacked monolith pieces on corundum substrates differing by the active component: front part – LaNiPt/CeZrO$_2$; rear part- Pt/CeZrO$_2$; total length 63 mm, feed preheat 400 °C.

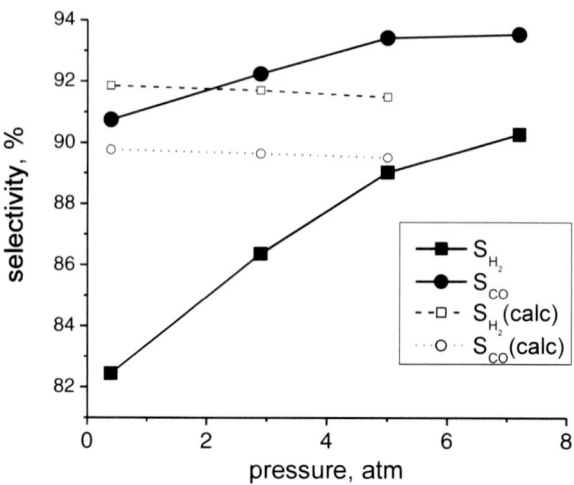

Figure 45. Pressure dependence of CO and H_2 selectivity. Experimental parameters as in Figure 44.

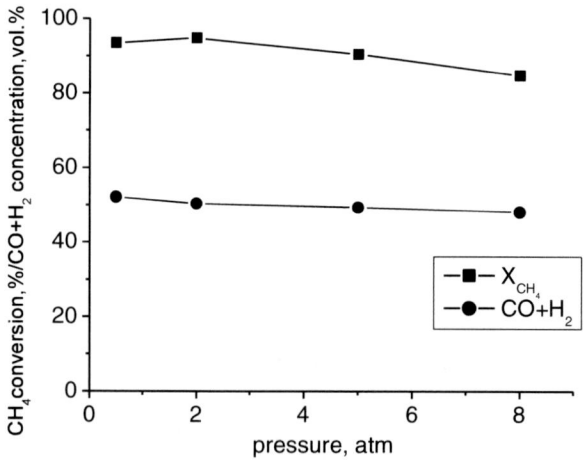

Figure 46. Pressure dependence of methane conversion and syngas concentration at contact time 0.12 s. Stacked monolithic layer. Feed 25 % CH_4 in air, inlet gas temperature 400°C.

In studied range of parameters, the pressure affects slightly the process parameters in both sets of experiments –at a constant contact time (STP) and a constant feed rate (contact time increases with pressure). All this may be explained by the nearly equilibrium product gas composition which is favored by the high catalyst temperature observed in studied pressure range.

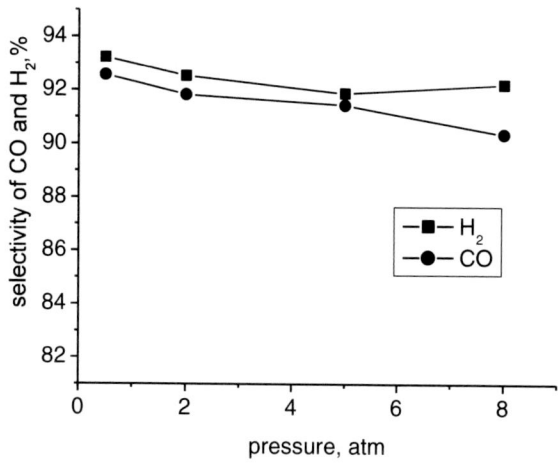

Figure 47. Pressure dependence of CO and H_2 selectivity at contact time 0.12 s (experimental parameters as in Figure 46).

5.1.5. Steam Reforming of Methane and Ethanol over Monolithic Catalysts

To study steam reforming (SR) of methane and ethanol (as well as other liquid oxygenate fuels), the monolith catalysts along with a front thermal shield were placed into a tubular stainless-steel reactor connected to the fuel/water evaporation unit with liquids supplied by the plunger pumps through nozzles. The temperature of the catalysts was measured by thermocouples located in the central channel at the exit of the monolith. The feed composition was 37 vol.%CH_4, H_2O balance. The contact time was varied in the range of 0.2-0.75 s by changing the feed flow rate, the oven temperature was kept in the range of 700-850°C.

Steam reforming of methane was studied over the catalyst LaNiPt/CeZrO$_2$ supported on the monolithic ZrO_2 substrate (Figure 12). At increasing contact time (Figure 48), concentration of H_2 and CO increases in the product gas, whereas the CO_2 concentration is practically unchanged. The H_2/CO ratio decreases with the increase of the contact time and temperature. Such dependencies are determined by the concurrent occurring of water gas shift reaction at temperatures ~ 600°C. Thus, the sum concentration of H_2 and CO attains 55 vol.% at 0.42 s contact time and catalyst temperature 620°C. There is a pronounced difference between the oven and the catalyst temperature due to heat consumption by the endothermic SR reaction. The higher is the oven temperatures, the higher is syngas content (Figure 49).

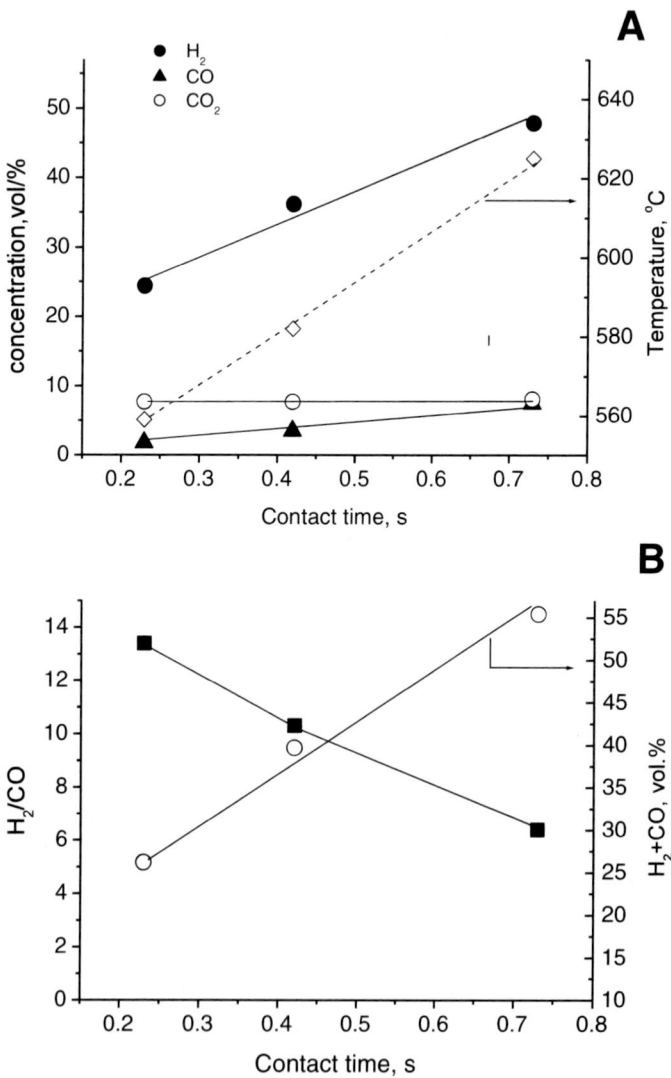

Figure 48. Catalyst temperature, H_2 and CO (A), H_2+CO concentration and H_2/CO ratio (B) in product gas versus contact time in steam reforming of methane over LaNiPt-CeZrO$_2$/zirconia honeycomb monolith. Feed 37 vol.% CH_4 in H_2O.

The presented results show that the reaction of methane steam reforming in realistic feeds with a small steam excess could be efficiently performed with the developed catalyst based on nanocrystalline oxides solid solutions promoted with a small amount of noble metals and supported on monolithic substrates.

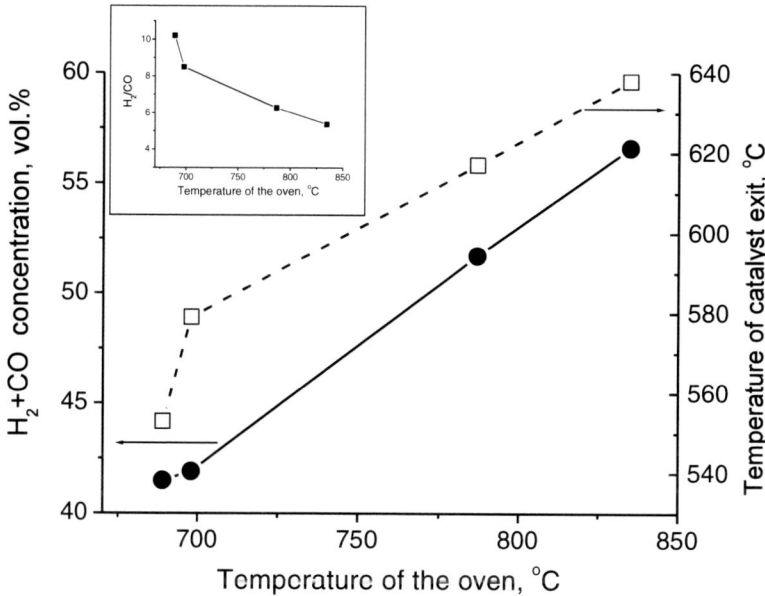

Figure 49. Dependence of the catalyst temperature, sum of H_2+CO concentration and H_2/CO ratio on the oven temperature in methane SR over the LaNiPt-CeZrO$_2$/ZrO$_2$ monolithic catalyst. Feed –37 vol.% CH$_4$ in H$_2$O, contact time – 0.42 s.

Steam reforming of ethanol was studied over catalysts comprised of 1%Ru/8%Ce$_{0.4}$Zr$_{0.4}$Sm$_{0.2}$O$_{2-x}$ supported on monolithic corundum substrate prepared via hydrothermal treatment route (Figure 11) or LaNiPt/CeZrO supported on extruded honeycomb zirconia substrate (Figure 12). The catalysts provide a high concentration of hydrogen and CO at rather short contact times and a small water excess (Figure 50-52). Main by-product is CH$_4$ which agrees with studies carried out for the catalyst fractions (Figure 7) Long-time testing (at least for twenty hours) shows a good stability of catalysts to coking (Figure 53).

Oxysteam reforming of acetone and sunflower oil. The monolithic catalysts with nanocomposite active components were shown to be quite efficient in oxysteam reforming of oxygenates providing a stable performance despite their well known coking ability (vide supra). Oxysteam reforming of acetone was studied on the monolithic catalyst LaNiPt/CeZrO/corundum substrate (Figure 12).

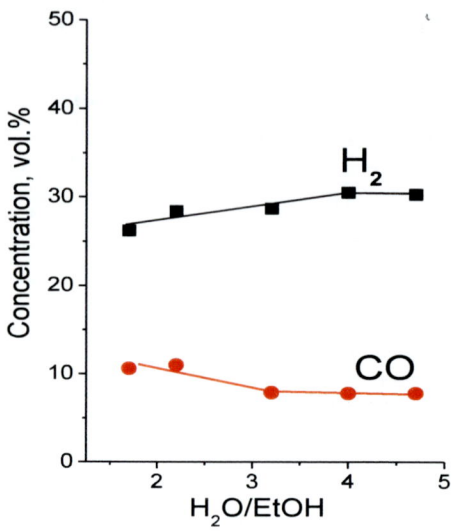

Figure 50. H_2 and CO concentration versus H_2O/C_2H_5OH ratio in steam reforming of ethanol over 1 wt.% Ru/8 wt.% $Ce_{0.4}Zr_{0.4}Sm_{0.2}O_2$/corundum monolith. T_{cat}=750 °C, contact time 0.1 s, 12 vol.% C_2H_5OH +H_2O, N_2 balance.

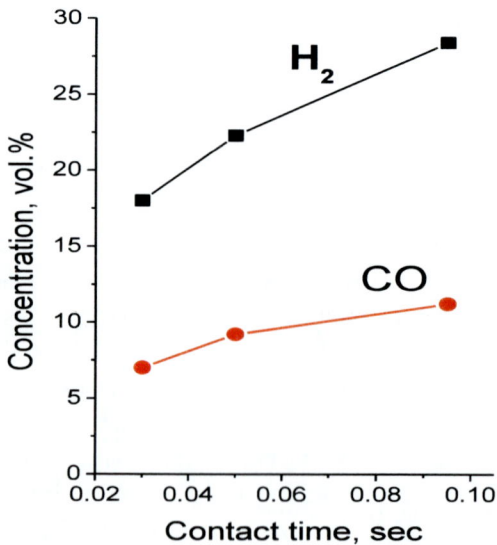

Figure 51. H_2 and CO concentration versus contact time in steam reforming of ethanol over 1 wt.% Ru/ 8 wt.% $Ce_{0.4}Zr_{0.4}Sm_{0.2}O_2$/corundum monolith. T_{cat} 750°C, H_2O/C_2H_5OH=2, 12 vol.% C_2H_5OH, N_2 balance.

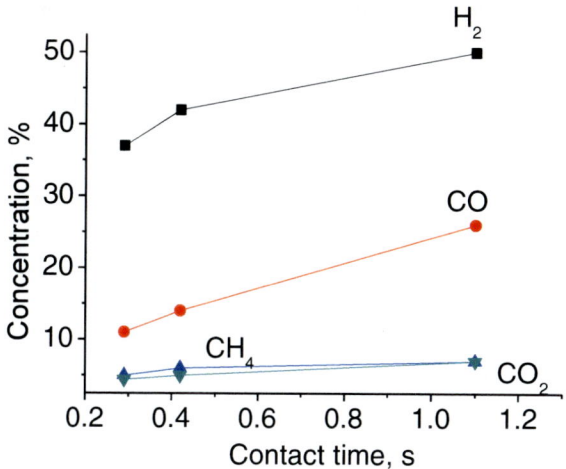

Figure 52. Effect of contact time on products concentration in exit feed for ethanol steam reforming on LaNiPt/Ce-Zr-O/honeycomb ZrO_2 substrate. Temperature 550 °C, H_2O +EtOH feed, H_2/EtOH = 1.7.

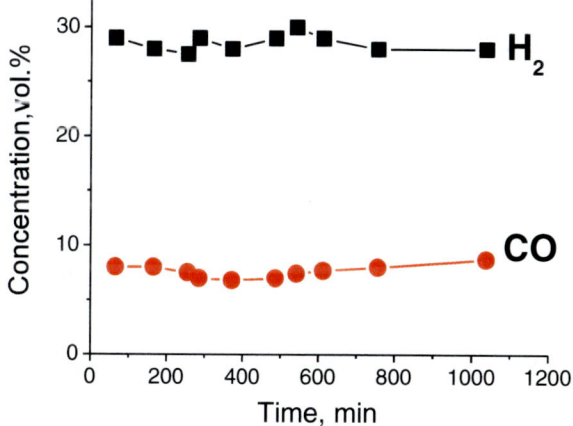

Figure 53. Long-time testing of 1%Ru8%$Ce_{0.4}Zr_{0.4}Sm_{0.2}O_{2-x}$/corundum monolith in steam reforming of ethanol. T_{cat}=750°C, H_2O/C_2H_5OH=3, contact time 0.1 s.

In case of oxysteam reforming of sunflower oil, which is more difficult to perform, a short piece of Fechraloy thick foil monolithic substrate (Figure 9c) with LaNiPt/CeZrO active component in the front part followed by a piece of monolithic catalyst on corundum substrate (Figure 12) with the same active component were loaded in the reactor. Typical results are shown in Figures 54 and 55.

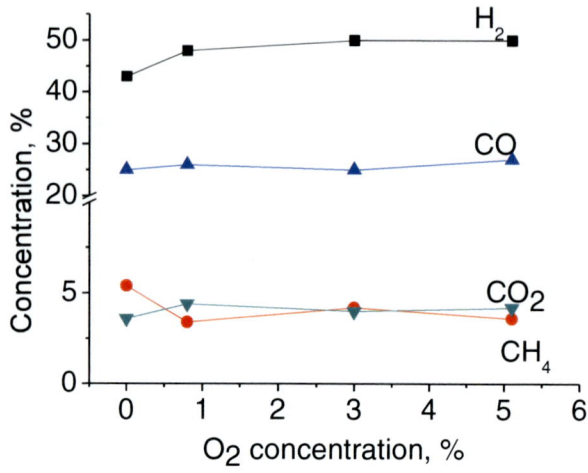

Figure 54. Effect of oxygen content in the feed on products concentration in acetone oxysteam reforming. Contact time 0.5 s, inlet feed 24% acetone + 48% H_2O + O_2, N_2 balance, T inlet 600 °C, T outlet 700 °C. Monolithic catalyst 7 %LaNi(Pt)O_3/10% Ce-Zr-O/corundum honeycomb substrate.

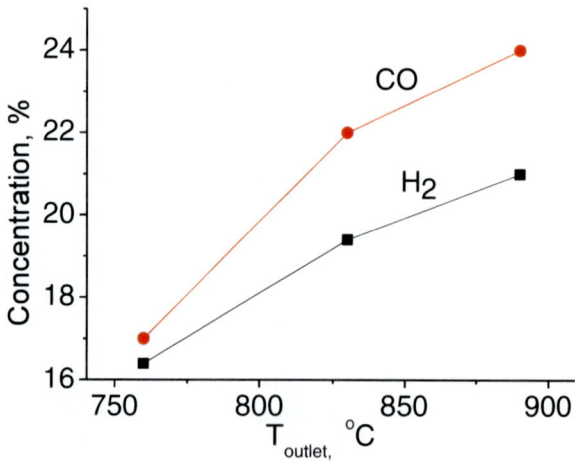

Figure 55. Effect of catalyst temperature on the product concentration in the oxysteam reforming of sunflower oil on stacked layer of monolithic catalysts (see text). Feed composition 0.7% of sunflower oil + 15% H_2O +20% O_2, N_2 balance, contact time 0.2 s.

Addition of oxygen to the feed provides required exothermicity in the reformation process and favors no-coke conditions (vide supra). The oxygen content in the feed does not affect the acetone oxysteam reforming, which is very attractive from the practical point of view. Main by-product CH_4 is formed due to acetone cracking on metal sites similar to the case of ethanol steam reforming.

Higher temperatures are required for sunflower as a fuel to provide a good yield of syngas as compared with more easily reformed ethanol or acetone. This was achieved by using the feed with a higher content of oxygen.

5.2. AUTOTHERMAL RADIAL-FLOW REACTOR WITH STRUCTURED CATALYSTS

The integrated reactor concepts are currently in the focus of intense research in reactor engineering. In these concepts, different unit operations such as heat exchange, distributed feed supply, or product separation are integrated in addition to the chemical reactors into one multifunctional apparatus by energy saving design [226].

A promising approach of the multifunctional reactor concepts employing recuperative heat exchange between the process streams and "entering and leaving energy and species along the full length of the apparatus" [226] was utilized in the radial-flow reactor configurations developed for the partial oxidation of natural gas to synthesis gas. It is known that the successful operation of a radial-flow reactor lies mainly in the details of the mechanical design, which is more complicated than that of a monolith reactor [227]. The autothermal radial-flow reactor system (see Figure 56, 57) includes a multiple cylindrical beds of catalysts including a stack of washers with milled channels covered by the layer of nanocomposite active component, layers of gauze catalyst with the microspherical catalyst bed between them, and a heat exchanger situated around the reactor as an outer shell, wherein the inlet air-methane mixture passes for preheating before entering into the central part of the reactor. Thus the feed is forced throughout the catalyst beds in the radial direction. Schematic drawing of the radial-flow reactor catalyst beds with approximate positions of the thermocouples is shown in Figure 56, 57. The radial flow monolith (a stack of the catalytic washers) 125 mm long was wrapped in the catalytic gauze, and then microspherical catalyst (LaNiPt (9 wt%)/Ce-Zr-Ox (~12 wt%)/γ-Al_2O_3, 1 mm spheres, $S_{sp.}$ 150 m^2/g) was gently loaded into a space between the gauze bed and an outer retention mesh. The catalytic arrangement was held between two flanges placed on the edges of the arrangement to ensure a good sealing and eliminate bypass flow of the reformate mixture.

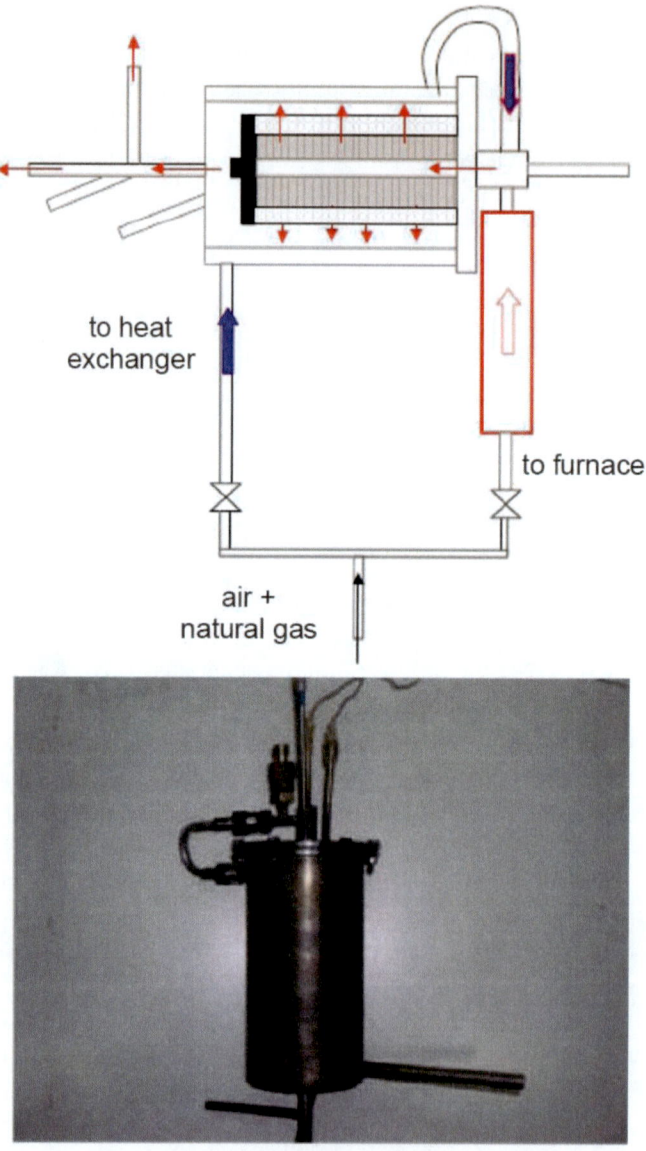

Figure 56. Autothermal radial-flow reactor configuration (see text for details).

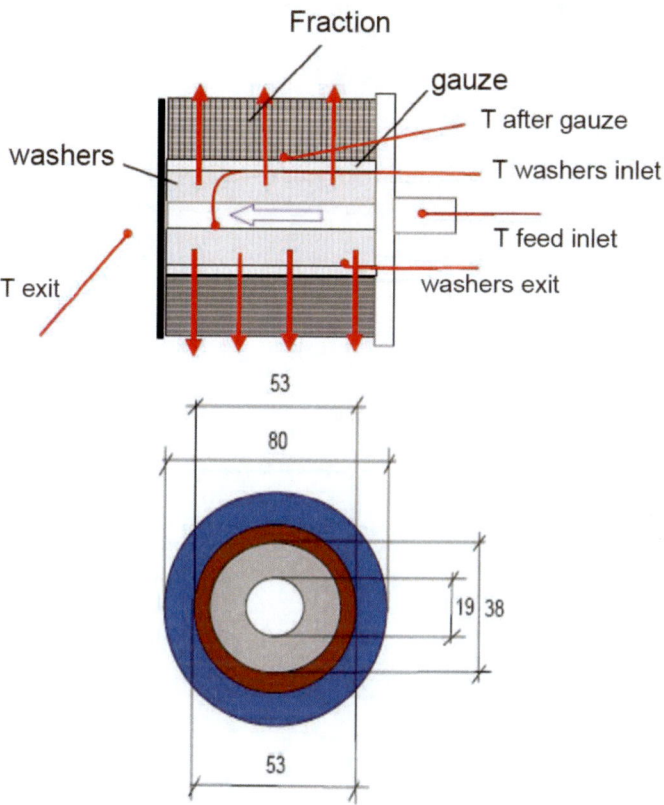

Figure 57. Schematic drawing of the radial-flow reactor catalyst beds.

During start-up period, an external heater was applied to heat up the reactor by air to ~450°C. Once an operational temperature was achieved, the feed (29 % natural gas in air) with a flow rate of 2 m³/h (STP) was fed at 425°C into the preheated reactor for enabling thus its quick start-up. In the stationary operation mode, an external heater was switched off. The inlet mixture at ambient temperature was introduced into the heat exchanger for preheating prior to entering the central part of the reactor. Further, the preheated feed was distributed along axis of the reactor, and flows in the radial direction across the catalyst beds.

Thus, the velocity of the gas is the highest at the inlet and then decreases as the gas fluid moves away from the inlet of the reactor. The reformed gas was collected into a plenum around the catalyst arrangement and exited from a single pipe. In this work this reactor was tested at the same natural gas - air ratio of about 0.33 in the feed with the flow rate varied from 4.3 to 14.4 m³/h.

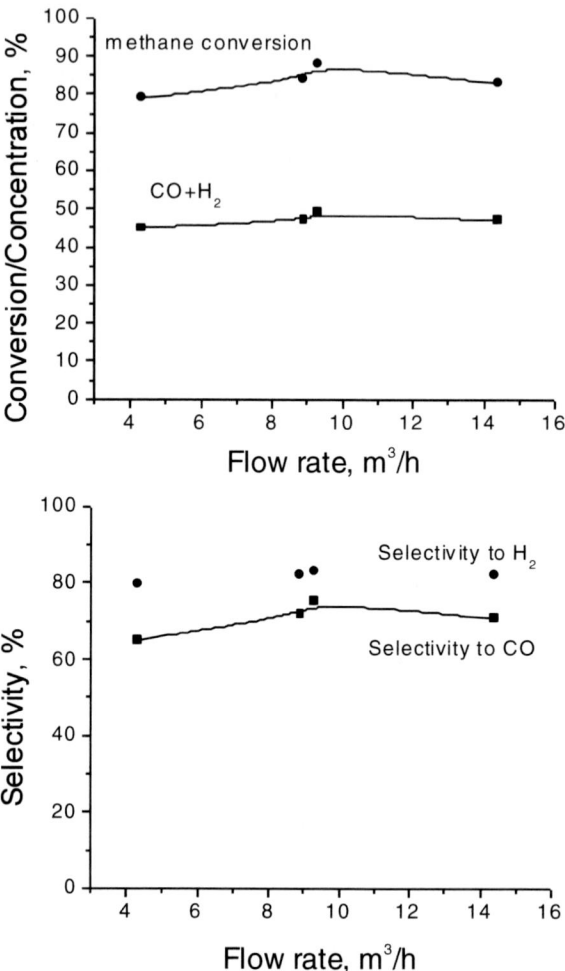

Figure 58. Methane conversion and syngas concentration (left), selectivity (right) versus flow rate for partial oxidation of natural gas in the autothermal radial-flow reactor.

Table 8 and Figure 58 show the results of testing the autothermal radial-flow reactor in the partial oxidation of natural gas (94 % CH_4, 3% C_2H_6, 2% C_3H_8, 0.4% C_4H_{10}, 0.1% CO_2, 0.5% N_2) with air. It demonstrated a stable performance without any external feed preheat at variation of the flow rate in studied range. The data in Table 8 suggest that, as expected, the general behavior of the methane partial oxidation process in terms of the temperature in the multiple catalyst-beds cylindrical architecture was similar to that observed for the axial flow monolith reactor.

Table 8. Autothermal radial-flow reactor performance results.
Feed: Natural gas-air mixture with $O_2/CH_4 = 0.768$

Flow rate, m^3/h (STP	Super-ficial velocity, m/s (STP	*Temperature, °C				Product species (experiment/thermodynamics), %v/v			
		Feed inlet	Washers inlet	Gauze bed outlet	Exit	H_2	CO	CO_2	CH_4
4.3	0.16	486	1146	832	648	30.1/ *27.4/ **32.6	14.3/ *12.1/ **17.0	3.74/ *5.0/ **2.1	3.85/ *3.7/ **0.1
8.9	0.34	473	1179	877	757	31.4/ *32.1/ **32.6	15.7/ *16.2/ **17.3	3.35/ *2.7/ *1.9	2.92/ *0.5/ **0.04
9.3	0.35	467	1111	868	737	31.8/ *31.7 **32.6	16.6/ *15.8/ **17.2	3.23/ *2.9/ **2.0	2.03/ *0.8/ **0.06
14.4	0.54	386	1014	854	503	31.2/ *14.1/ **32.6	16.1/ *2.8/ **17.2	3.75/ *10.5/ **2.0	3.56/ *11.7/ **0.06

Notes: Positions of the thermocouples are shown in Figure 57.
*Equilibrium at the exit temperature.
**Equilibrium at the outlet temperature of the gauze bed.

Temperature reaches the highest value at the very front of the radial flow monolith comprised of stacked washers. The temperature profile across the radial-flow reactor catalyst beds declines towards the exit of the reactor. It is important that the radial flow arrangement shows a high stability in the temperature in both the radial flow monolith and gauze catalyst bed at the feed volume flow rate being varied, due to high thermal conducting properties of metal substrates and minimization of heat loss. This apparently helps to keep constant the natural gas conversion and syngas selectivity independent of the flow rate (Figure 58).

The reversible nature of these reactions imposes a limit determined by thermodynamic equilibria, on the conversion and yields of CO and H_2. Thermodynamic equilibrium compositions (on the dry basis, same as that measured by the GC) under the experimental conditions at the reactor exit temperature (tagged as * in Table 8) and the gauze bed temperature (tagged as **) suggest that CO and H_2 concentrations in the product gas are most likely determined by equilibria at the temperature in the microspherical catalyst bed at a position close to the gauze bed. Unfortunately, no data was available from this zone of the radial flow reactor.

The most prominent deviation from the measured concentrations appeared in the thermodynamic predictions calculated at the reactor exit temperature for the case of 14.4 m³/h flow rate. The exit temperature is rather low in this case due to a high heat transfer in the recuperative heat exchange between the cold feed stream and hot product gas. However, the syngas production rate is high. Seemingly, direct quenching of the product gas occurs by the cooling in the microspherical catalyst bed through the heat exchange with the inlet feed.

5.3. SYNGAS GENERATION BY CATALYTIC PARTIAL OXIDATION OF N-DECANE

Two types of structured catalytic reactors with the catalysts based upon thermal conducting metallic substrates (fecraloy) have been successfully tested in partial oxidation of n-decane to synthesis gas:

(1) metallic monolith of the Arkhimed spiral type (Figure 9).
(2) structured catalytic package made of stacked fechraloy foil sheets and gauzes (Figure 59, 60) protected by corundum layer supported by blast dusting technique [122, 228, 229]. The active component was comprised of Pt (1.6 wt.%)- or Rh (0.4 wt.%) + Pt (0.8 wt.%)- promoted $LaNiO_3$ (4 wt.%)/$La_{0.2}Zr_{0.4}Ce_{0.4}O_x$ (5 wt.%)/alumina [124, 229].

A blank alumina honeycomb monolith was placed before the catalyst as a thermal shield. There were two operating variable in the experiments: contact time (residence time) and oxygen-to-fuel ratio expressed as the molar ratio $O_2|C$. Concentration of n-decane in the feed was varied in the range of 2.86-4.94 %. The total feed flow rates were adjusted to provide the contact times in the range of 0.27-13 ms (STP).

Figure 61 shows syngas yield (left) and composition of the product gas (right) during partial oxidation of n-decane in air as a function of O_2/C ratio in the feed. The maximum of syngas yield was found at O_2/C =0.65. When $O_2/C>0.65$, syngas yield (first of all, hydrogen) decreased due to combustion, but for all that higher oxygen content in the feed has more pronounced effect on the concentration of hydrogen then CO. Behavior of the other major products indicates that the deep oxidation route is getting more visible at further increasing oxygen –to carbon ratio. However, the feed flow rate varied in wide limits affects rather the catalyst temperature than product gas composition.

Thus, decrease in the contact time in 3 times (from 12 ms to 4 ms) resulted in decrease of synthesis gas yield only by 7%. Maximal concentrations of desired products, H_2 and CO, were detected at the lower limit of flow rate (1.8 m³/h).

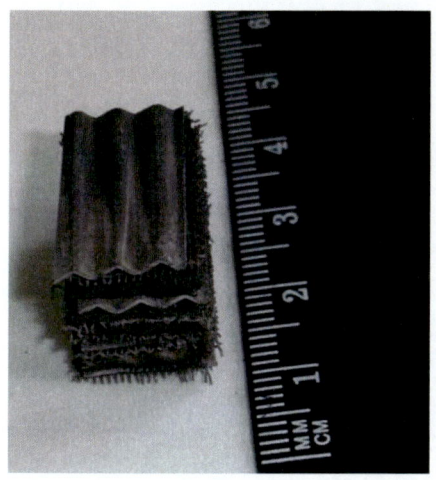

Figure 59. Structured catalytic package comprised of sheets and gauzes.

Figure 60. View of a reactor with the catalytic package for studies of n-decane partial oxidation at contact times 0.27-2 ms.

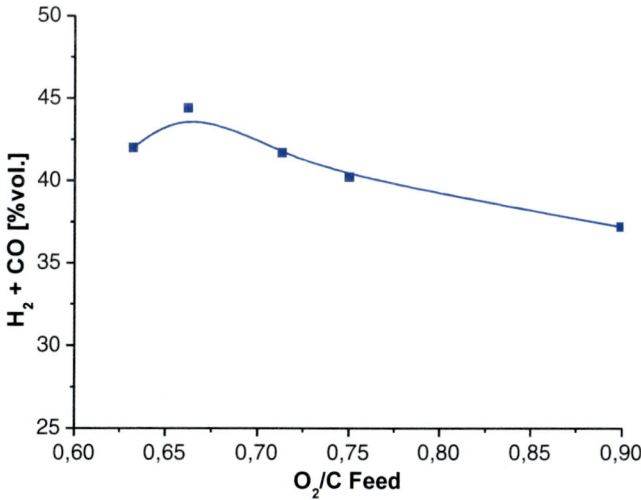

Figure 61. Continued on next page.

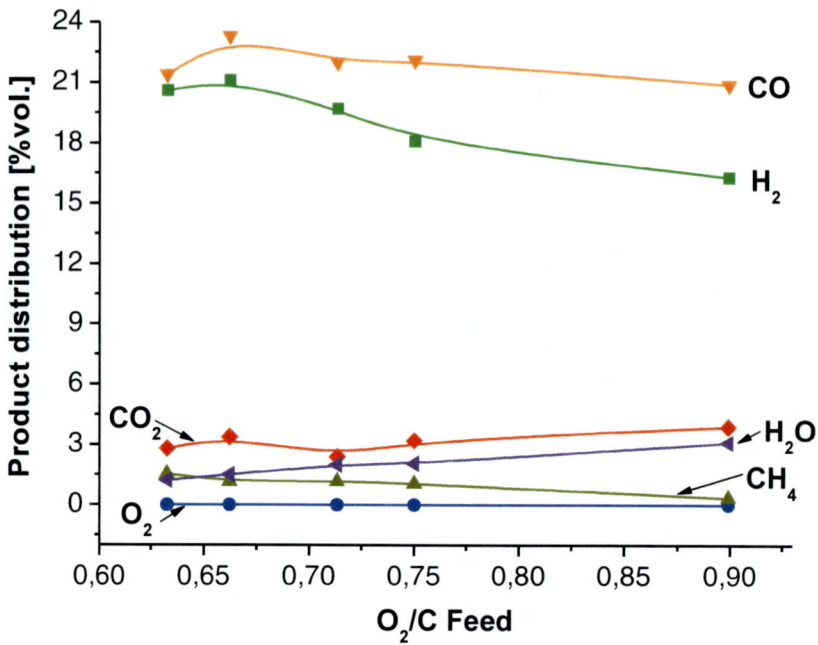

Figure 61. Syngas content (left) and composition of the product gas (right) in partial oxidation of n-decane in air as a function of O_2/C ratio in the feed. Monolithic catalyst on FeCrAl thick foil spiral substrate. Feed n-decane +20% O_2 + N_2 balance, contact time 7 ms, T_{exit} 950 °C.

The catalyst temperature increases with the flow rate as expected (Figure 62) which helps to maintain a high syngas yield even at very short contact times.

As contact time increased, the backside catalyst temperature declined steadily (Figure 62, right), thus indicating that contribution of endothermic routes became more important. The temperature in front of the catalytic monolith is an intricate function of the interplay of chemistry and heat-mass transfer. In partial oxidation of decane at very short (0.27-2 ms) contact times on package comprised of stacked fechraloy sheets and gauzes, to ensure a stable performance in runs with a low exothermicity (at O_2/C ratio <0.5), a resistive heater (T~ 300 °C) coiled around this package was used to compensate the heat loss.

63 and 64 show main performance features for two structured catalytic packages with active components comprised of Pt-Ni/La-Ce-Zr-O with or without Rh addition. Doping by Rh allows to run the process autothermally with a lower O_2/C ratio and in a broader range of contact times as well as to obtain a higher syngas content.

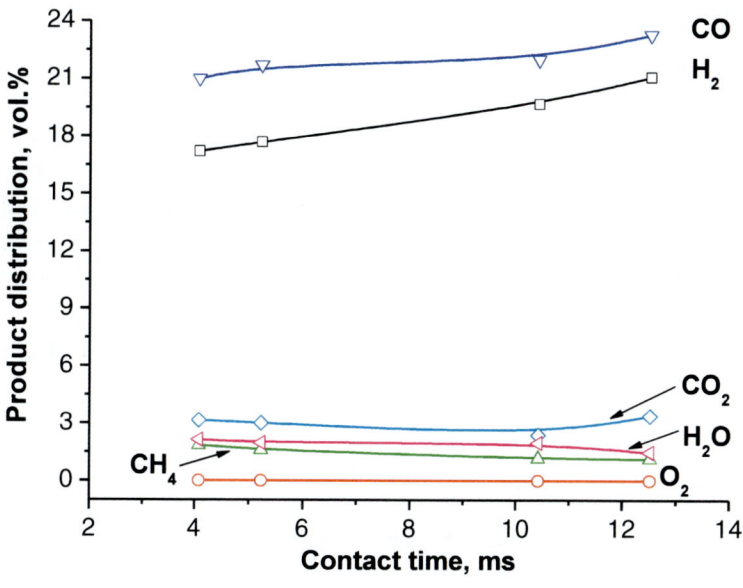

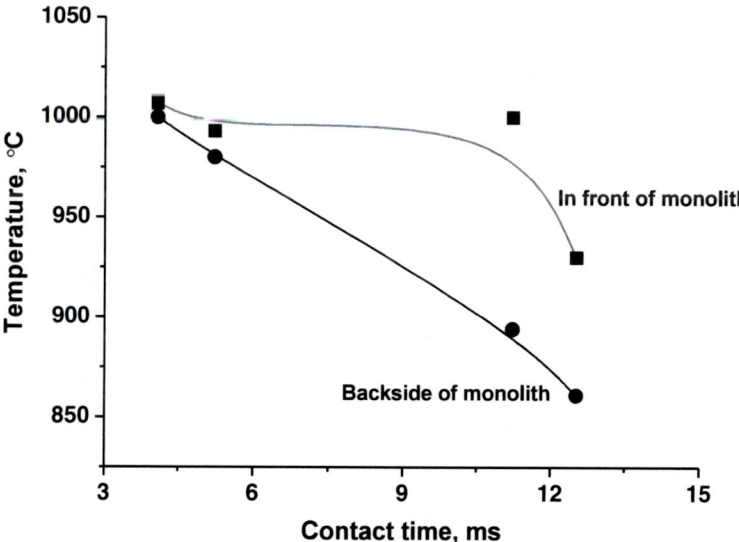

Figure 62. Steady-state reaction temperatures in the metallic monolith of the Arkhimed spiral type versus contact time (STP) in partial oxidation of n-decane. Feed composition: decane (2.86 vol.%):O_2 (20.4 vol.%): N_2 (75.8 vol.%). Flow rate 1.8 - 4.9 m^3/h.

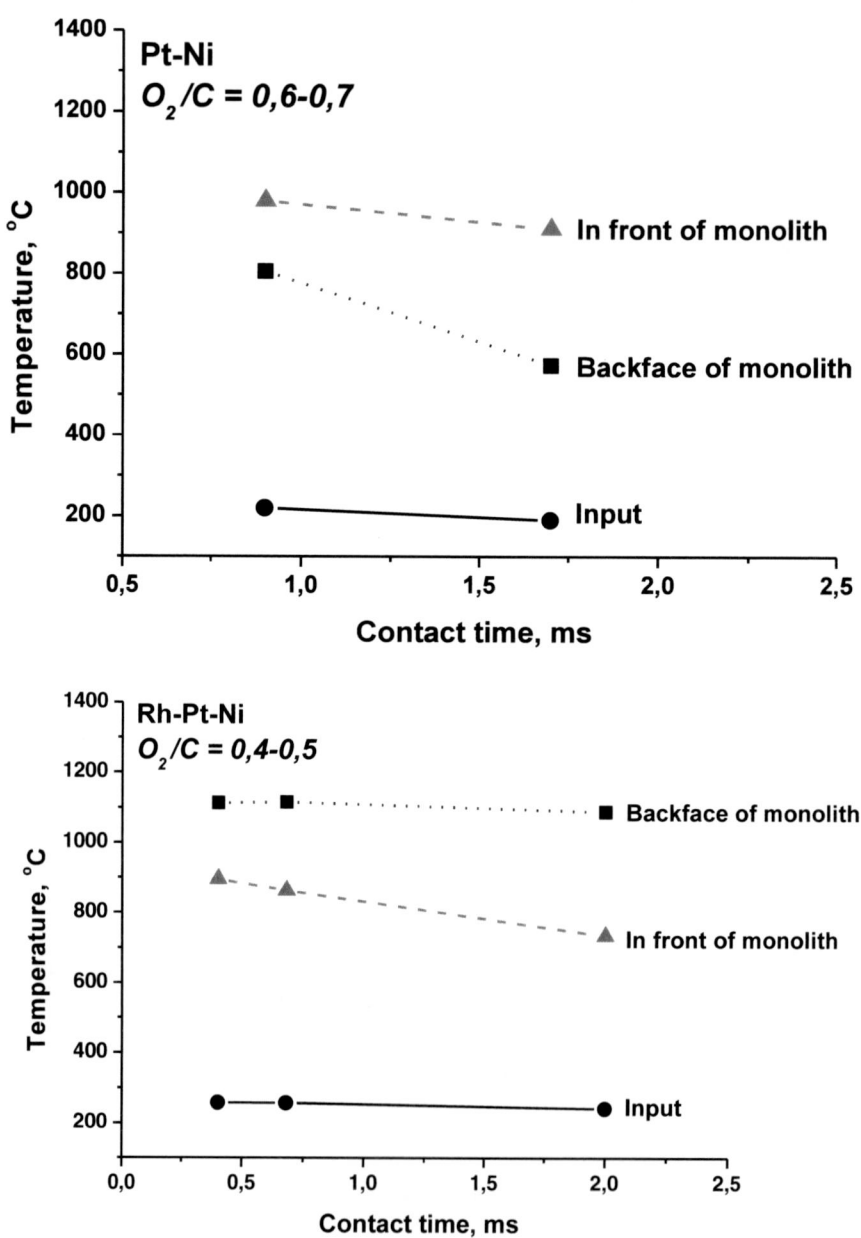

Figure 63. Effect of contact time on the reaction temperature in partial oxidation of decane on structured catalyst packages. Feed decane + air, O_2/C in the range of 0.4-0.7.

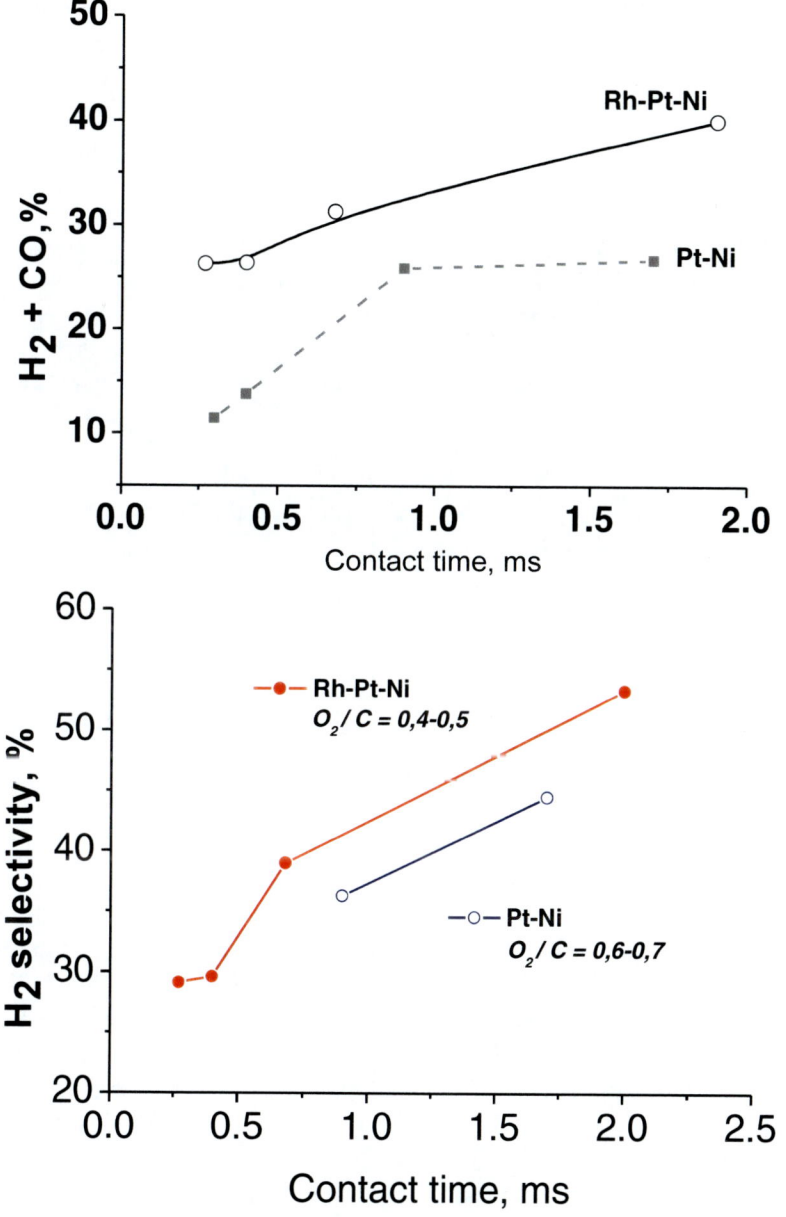

Figure 64. Syngas content (left) and H_2 selectivity (right) versus contact time during partial oxidation of decane over structured catalysts packages.

Figures Performance was stable for both types of active components for at least for 10 h time-on-stream. As follows from SEM images of catalyst surface before and after reaction (Figure 65), there is no carbon deposition or active component sintering/detachment. The porous washcoat layer seems to form some sort of rhodium/alumina composite layer which does not allow the rhodium to sinter or vaporize.

Figure 65. SEM images Rh-Pt-Ni/La-Ce-Zr-O catalyst surface: A – before reaction; B – after reaction.

5.4. SYNGAS GENERATION BY CATALYTIC REFORMING OF INDUSTRIAL MINERAL OIL

Mineral oil is a very complex mixture of low and high (C_{15}–C_{50}) molecular weight aliphatic and aromatic hydrocarbons, additives, metals, and various organic and inorganic compounds [230]. The chemical composition of the lubricating oil varies widely and depends on the original crude oil, the refining processes, an engine that utilizes the oil and the additives in the original oil. Most of the publications on the steam reforming and cracking of oil are related to bio-oil and tar derived from pyrolysis of biomass and wood. There are just a few scientific publications about gasification of mineral oil (C_{15}-C_{50}), e.g. [231-233].

Autothermal reforming reaction of industrial mineral oil in the presence of air and steam over Rh-Pt-Ni/La-Ce-Zr-O mixed oxide catalytic composition supported on FeCrAl monolith substrate (vide supra) has been studied at the

Boreskov Institute of Catalysis. Elemental composition of mineral oil (Table 9) was analyzed with "ELEMENTAR vario EL III" and "Horiba SLFA-20"analyzers.

Table 9. Elemental composition of the lubricated mineral oil

C	H	N	S
84.92±0.12 %	15.22±0.17 %	0.04 %	1263±9 ppm

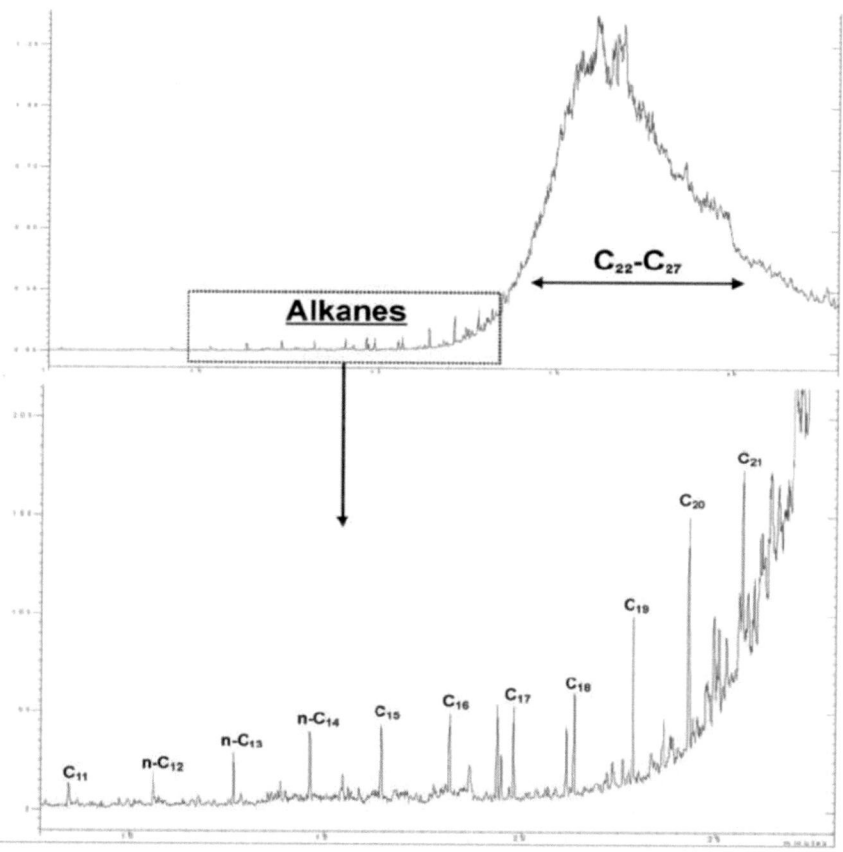

Figure 66. Chromato-mass spectrum of mineral oil.

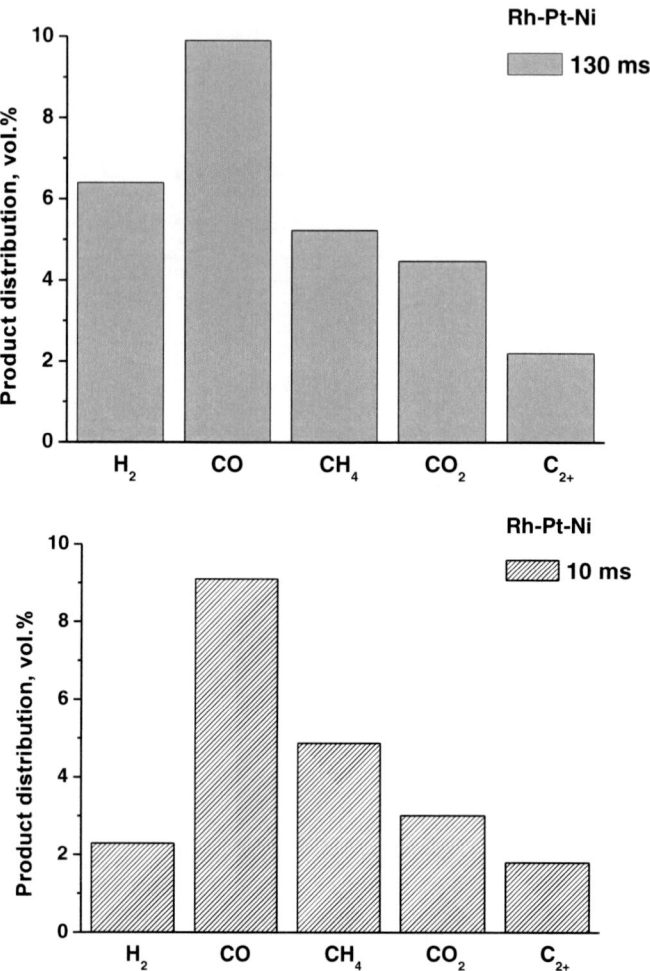

Figure 67. Product distribution in autothermal reforming of mineral oil at 650°C on structured catalyst based on Rh-Pt-Ni/La-Ce-Zr-O active component supported on (a) Arkhimed spiral type substrate, feed 9.9.% O_2 +49.51 % H_2O + 3.37.% mineral oil, N_2 balance, contact time 130 ms (STP). (b) package of fechraloy sheets and gauzes substrates, feed 10.1.% O_2 + 44.95 %H_2O + 6.81 % mineral oil, N_2 balance, contact time 10 ms (STP).

Some results of activity tests are given in Figure 67. Even at 650°C and at short contact times, syngas is present in converted stream along with products of deep oxidation (CO_2) and cracking (CH_4, C_{2+}). Longer contact times give higher H_2 content.

Molar ratio C/H of 2.15 was determined. The component content of the fraction according to chromatographic/mass spectrometry (Varian mass-spectrometer "Saturn-2000" detector equipped with ion trap) is shown in Figure 66. The main fraction was determined to be high molecular weight C_{22}-C_{27} alkanes. The reforming reaction exothermicity is too low to be carried out at the autothermal mode. Therefore, the catalyst temperatures of 650 - 1000°C were maintained by conductively transferring heat from an electric heater to the catalyst. Determination of such products as H_2, O_2, N_2, CH_4, CO, CO_2, H_2O and C_x –gas phase hydrocarbons was carried out with gas chromatography. Composition of all separated hydrocarbon products including liquids condensed in cooled traps was also detected by using chromato-mass spectrometry.

A lot of other products including olefins, aromatics and oxygenates were revealed by both gas-phase and condensed phase analysis with chromato-mass spectrometry (Figure 68-70). This indicates a complex character of processes (oxidation, cracking, dehydrogenation, condensation, etc) occurring both on the catalysts and in the gas phase in the course of mineral oil oxysteam reforming. Apparently a bigger part of these intermediate compounds is transformed into syngas at higher temperatures, which is reflected in the increase of syngas content with temperature (Figure 71).

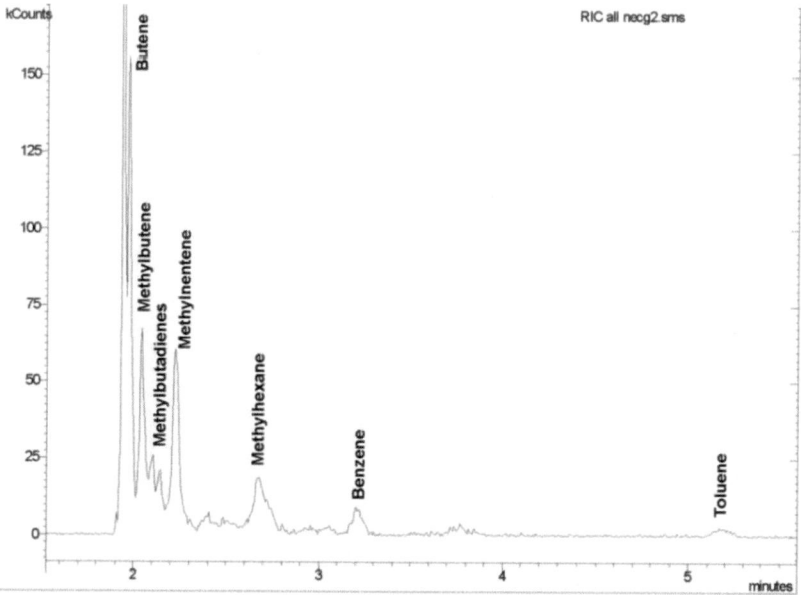

Figure 68. Chromato-mass spectrum of gas products (see Figure 67a for experimental conditions).

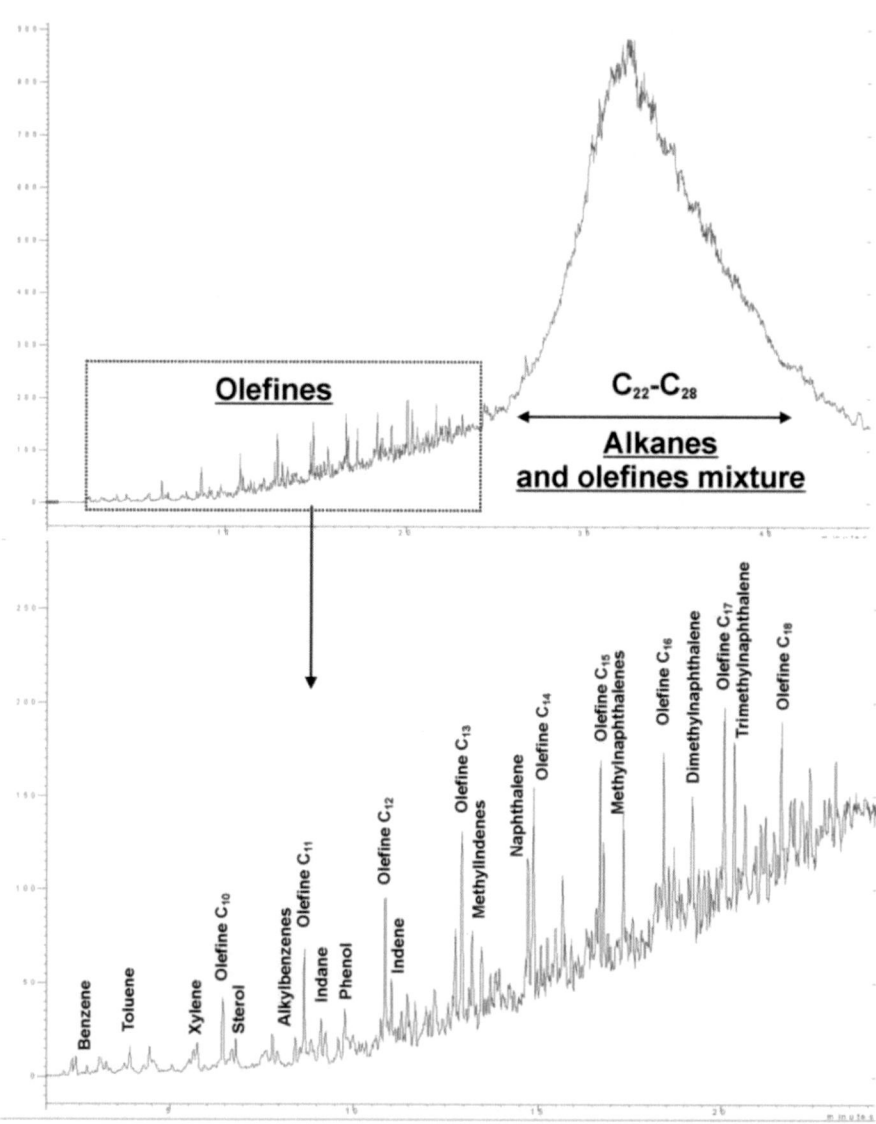

Figure 69. Chromato-mass spectrum of products organic phase condensed in a trap (see Figure 67a for experimental conditions).

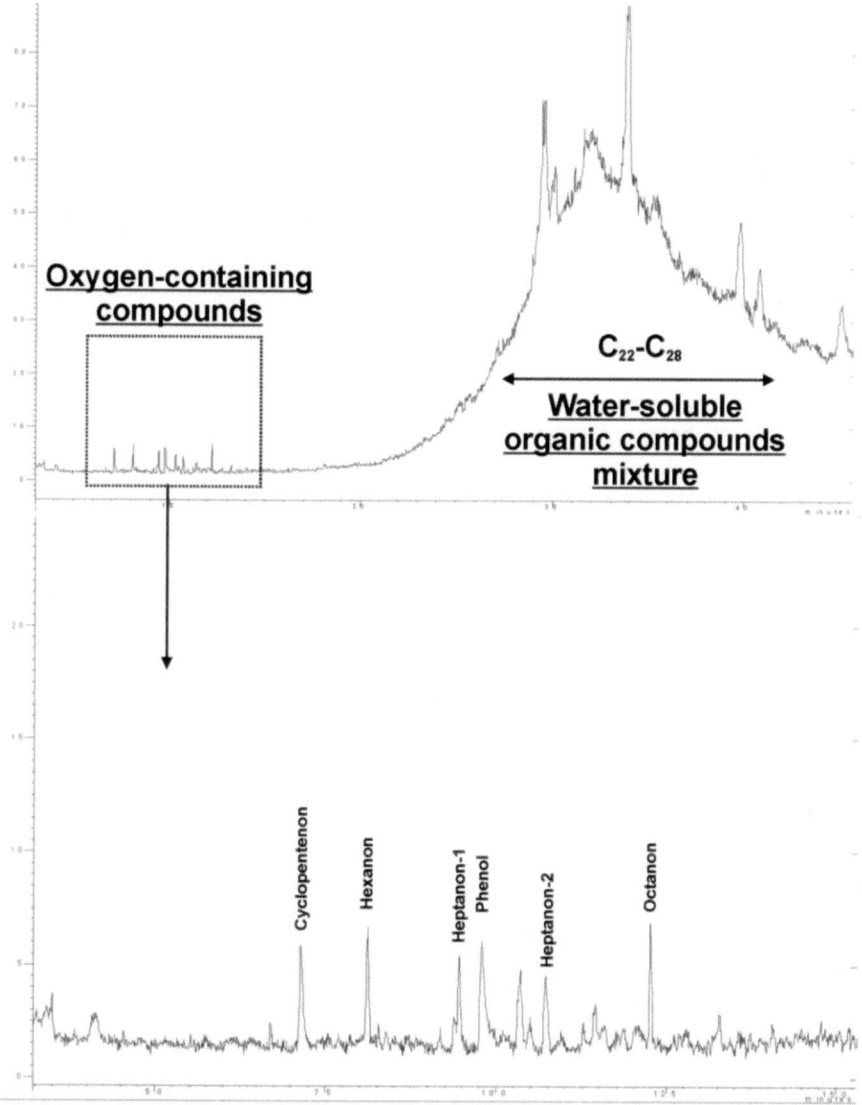

Figure 70. Chromato-mass spectrum of water-soluble phase of products condensed in a trap.

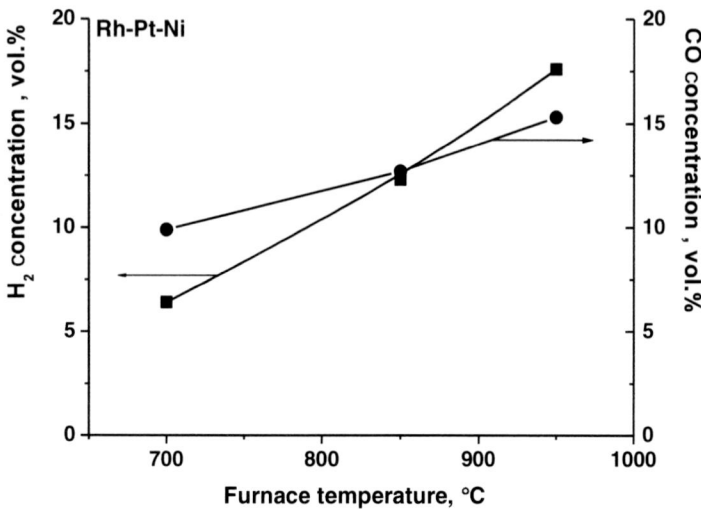

Figure 71. H_2 and CO concentration in the product gas as a function of the furnace temperature for the oxysteam reforming of mineral oil on monolithic catalyst based on Arkhimed spiral.

5.5. SYNGAS GENERATION BY FUEL REFORMING IN MICROREACTORS

Reforming in microchannel reactors enables extremely high mass and heat transport rates. Therefore, small and compact devices with small energy losses can be designed. This provides a good dynamic behaviour and a high conversion efficiency for reforming of hydrocarbon fuels and oxygenates where the complex coupling of catalysis, heat transfer and reactor design takes place. The intensive investigations of reforming operation conditions and catalytic systems in microchannel devices have been carried out at Boreskov Institute of Catalysis. Different types of microchannel devises have been developed [234-240].

Partial oxidation of methane. Figure 72 shows a microchannel catalytic reactor for partial oxidation of methane at a scale that would support 0.5 kW electrical output from a fuel cell [236]. Four such units were placed in a fuel processor, which enabled to produce 1.5 m^3/h (STP) of syngas. The fuel processor was operated at the consumption of 0.6 m^3/h methane with conversion approached 98.5 % and hydrogen selectivity of about 80 % in excess of air of 20% over stoichiometric value [239, 240].

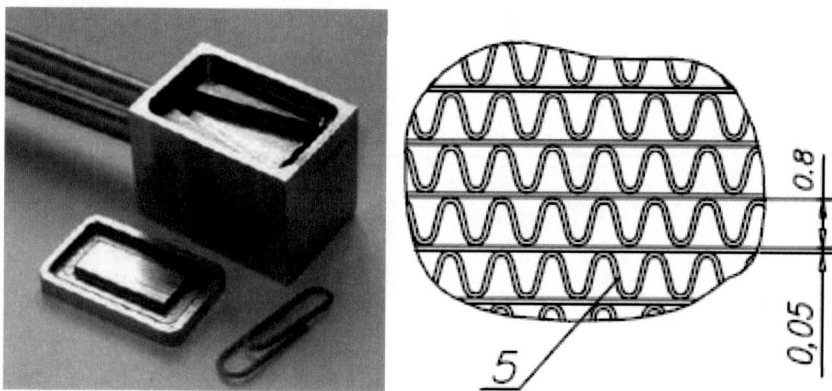

Figure 72. Microreactor for partial oxidation of methane, dimensions: 36x24x26 mm (left); microchannel catalytic stack comprised of corrugated and flat plates with dimensions of 20x20x0.8 mm.

In the microchannel unit (Figure 72), methane-air flow enters an upper trapezoidal chamber providing uniform distribution of gas flow through all catalytic microchannels in the reaction chamber. The reaction chamber consists of a set of 10 pairs of flat and corrugated sheets of fechraloy foil of 50 μm thickness, stacked to each other as shown in Figure 72 (right). Payload volume is about 4000 mm^3. A thin (~ 5-10 microns) protective corundum sublayer was supported on the fecraloy foil by blast dusting technique [229] followed by supporting active component LaNiPt/LaCeZrO$_x$ in amount of ca. 80 mg per one sheet as described in [228]. Results of the reactor test study are shown in Figure 73. At a feed flow rate of 350 ml/s (STP) corresponding to ca. 11 ms contact time, the product gas composition is (in mol.%): 31.6 H$_2$, 17.1 CO, 4.6 CH$_4$, 1.9 CO$_2$, 44.5 N$_2$ (dry gas), which corresponds to the adiabatic temperature of 804°C. A weak dependence of CH$_4$ conversion and syngas selectivity on space velocity provides nearly linear dependence of syngas productivity on the feed rate.

Methanol steam reforming. Methanol is widely discussed as an attractive source of hydrogen due to its high energy density, low cost and easy transportation. In addition, the reaction of methanol steam reforming does not require high temperatures that results in an easy temperature control. The performance in the steam reforming of methanol has been studied for two types of microreactors (Figure 74) differing by the arrangement of gas flows in the catalytic plates (cylindrical and rectangular) [238-240]. The microchannel plates (Figure 10) made of stainless steel foil (a), aluminum (b), and porous nickel (c) were manufactured by using techniques of laser etching, electrochemical etching, and plastic pressing, respectively. The cylindrical stainless steel based plates were in the form of disks of 29 mm diameter with 370 μm thickness for stainless steel and 25O μm thickness for aluminum alloy. A CuO/ZnO (40:60 mol/mol) catalyst–pseudoboehmite binder mixture was pressed in the cavities.

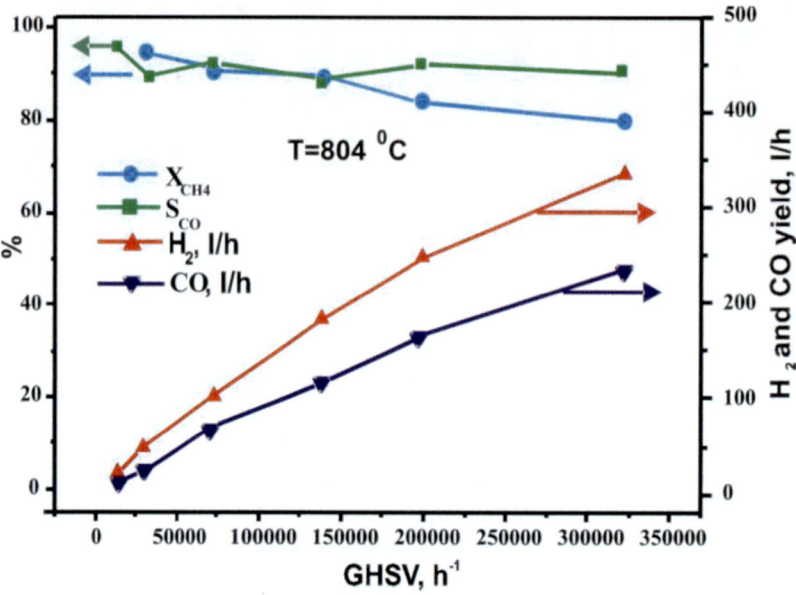

Figure 73. Methane conversion (X), CO selectivity (S), and yield of hydrogen and CO versus space velocity. Feed 29.4 % CH$_4$ in air.

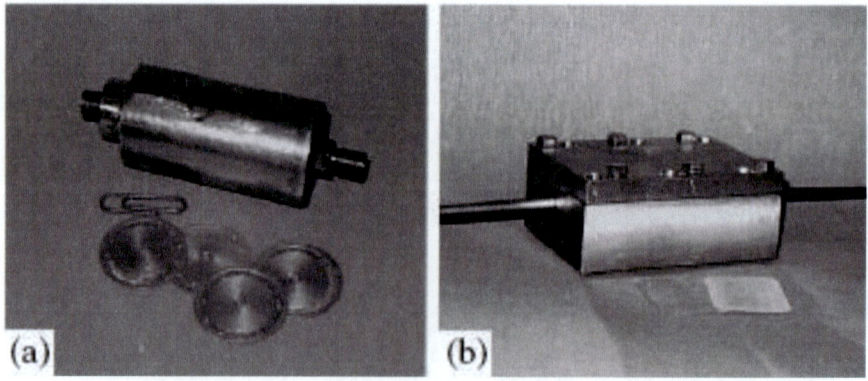

Figure 74. Microcreactors and corresponding microchannel catalytic plates of cylindrical (a) and rectangular (b) architecture.

The rectangular catalytic plates (Figure 10c) were made of porous nickel (ZAO Novomet-Perm) with porosity of 0.8. An original porous nickel plate of 2–3 mm thickness was loaded with the same catalyst–binder powder (70:30 w/w) by using ultrasonic dispersion in ethanol. After drying in air, the plate was cold-pressed under 250 atm with a special die. Fabricated rectangular plates with dimension of 30x40 mm and of 0.5 mm thickness have straight channels of desired geometry (Figure 10b).

Three types of microreactors were used in the work. A rectangular microreactor (RMR) was made of aluminum alloy, a cylindrical type microreactor (CMR) was made of stainless steel, and a tubular microreactor (TMR) was designed as a stainless steel tube.

In the cylindrical microreactor (CMR) of 30 mm diameter, and 60 mm long (Figure 74a), the reaction chamber contains 130 catalytic disk-type plates. The water-methanol (1:1 molar ratio) gas mixture flows from the microreactor center to the periphery along the spiral channel in the radial direction of the plates, going ultimately through holes in the catalytic plates to the output stream.

The unit at the feed flow of 150 ml/h and the reaction temperature of 260°C afforded 80 % conversion of methanol with hydrogen yield of 130 l/h. In terms of volumetric capacity of the microreactor, the hydrogen yield was 5.4 l/(cm^3 h). In the rectangular microreactor (RMR), the highest hydrogen capacity was achieved for the catalytic plates having two channels with a cross section of 0.2x 10 mm. It was found that the specific hydrogen capacity on the catalyst weight basis was more than three times higher for cylindrical architecture than that for the rectangular microreactor (Figure 75).

This is because of internal diffusion limitations for the porous nickel plate with a big loading of catalyst, in contrast to thin catalytic layer in the cylindrical architecture. However, on the volumetric basis, the specific capacity of the rectangular microreactor was nearly three times as high as that of the cylindrical microreactor (Figure 75).

The reaction of the methanol steam reforming is accompanied by the reverse water gas shift reaction producing CO which is harmful for PEM fuel cell operation. The CO outlet concentration increases with increasing the methanol conversion (Figure 76). Compared to the conventional fixed-bed TMR filled with the catalyst particles of 0.25–0.5mm in size, the microreactor with the catalyst particles less 50 microns yields much less CO. This is explained by the fact that large particles favor reverse water gas shift due to a higher concentration of both hydrogen and carbon dioxide in the catalyst grain center.

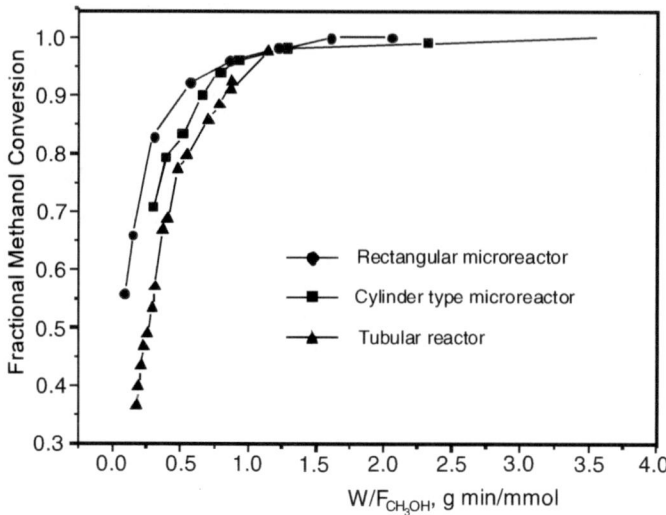

Figure 75. Dependence of methanol conversion *x* vs. *W/F*, where *W* (g) is the catalyst weight and *F* (mol/min) is methanol molar flow at the microreactor inlet. 266.7 °C. CMR-squares, RMR-circles, TMR-triangles.

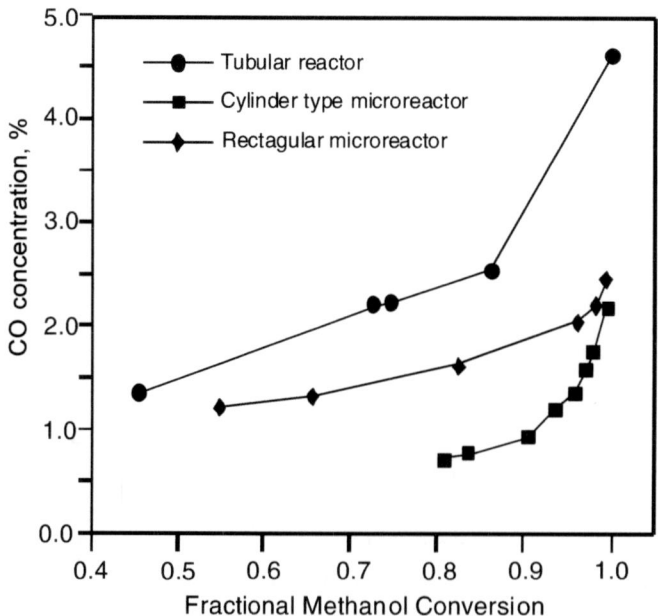

Figure 76. CO content in the exit stream vs. methanol conversion in the microreactors. 266.7 °C.

Since traditional CuO/ZnO catalyst is known to deactivate at enhanced temperatures due to sintering, a new type of highly active ZnO/TiO_2 catalyst was developed [240]. This catalyst was shown to be a solid solution of Zn cations in the anatase structure. Study of Zn/TiO_2 catalyst supported on the microchannel plates showed that the catalyst exhibited sufficient activity in the SMR processes and outperformed the known copper catalysts in some cases. The maximal catalyst activity was observed on the foam copper microchannel plate MCP2 (Figure 77). Thus, a microreactor with a total volume of 12 cm^3 at 450°C and the 80% methanol conversion produced 78.6 l/h of hydrogen,. The hydrogen production in the microreactor with the microchannel plates of foam nickel and corrugated brass foil (MCP1 and MCP3) was 1.6 and 2.4 times respectively lower than that with foamed copper plates (MCP2). Specific hydrogen production (relative to the catalyst mass) for MCP1 and MCP3 was 1.6 and 1.2 times respectively lower than that for MCP2 due to a lower catalyst mass in MCP3. Thus, from the practical standpoint, foam copper is the best material for microchannel plates used in the above microreactor followed by the nickel foam.

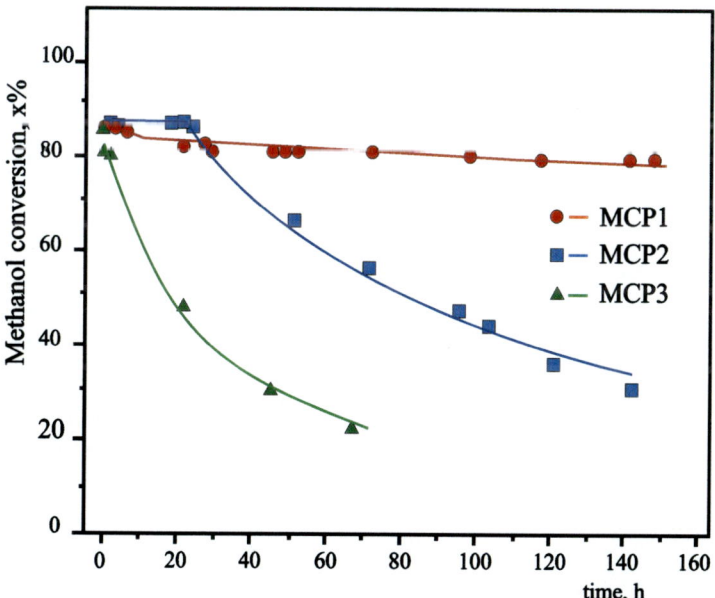

Figure 77. Performance variation with time-on-stream at 400 °C for Zn/TiO_2 catalyst supported on different microchannel plates. H_2O/methanol=1.1. For MCP1 (Cu foam) and MCP2 Ni foam) F=8.78 mmol/min, MCP3 (corrugated brass) F= 1.77 mmol/min.

Long-term tests (Figure 77) showed that stability of Zn/TiO_2 catalyst supported on the different microchannel plates in the SMR process depends on the material of the microchannel plate used. The highest operation stability of the catalyst was observed for the foam copper microchannel plates. Thus, during a 160-hour continuous operation of a microreactor equipped with the foam copper microchannel plates, the initial methanol conversion of 80% decreased by 8%. During 500-hour tests, the methanol conversion decreased by 32%. The main reason of the catalyst deactivation is sintering of zinc microparticles which are responsible for the catalyst activity. Hence, the rate of the catalyst deactivation directly depends on heat conductivity of a material of the microchannel plates. Activity of deactivated catalyst can be restored by annealing it in air. An increase in the concentration of water and addition of oxygen in the inlet mixture improves the catalyst activity and reduces its deactivation.

Chapter 6

UNBIDDEN PHENOMENA
IN THE REFORMING REACTOR

It has been previously shown [189] that non-catalytic pre-reforming of various hydrocarbons may occur in the reactor in front of the monolithic catalyst, affecting both conversion and selectivity of the reforming reaction. It was found that for a given feed composition, the superficial velocity in the reactor was the main parameter regulating the intensity of the pre-reforming process. Figure 78 depicts a tentative scenario of the temperature evolution in the monolith reactor at too low superficial velocity of gasoline- air feed mixture. In the experimental run the reactor was charged with a package of catalytic and two unloaded (blank) monoliths. At the linear velocity of 0.26 m(STP)/s, a uniform reaction front moved slowly to the inlet. Decline of the catalyst temperature then occurs followed by the ultimate cessation of the reforming process.

The pre-reforming phenomenon has been attributed to the thermal reactivity of the feed rather than ability of the catalyst to coke resistance. As it was discussed already, heavy hydrocarbon fuels and bio-fuels easily form coke due to thermal cracking, especially in the oxygen-containing mixtures. Figure 79 shows a view of the experimental quartz reactor after a few hours run of the partial oxidation of diesel fuel.

An additional factor in the process performance is the choice of materials for reactor design.. A stainless steel reactor loaded with metallic monoliths was used in the experimental run illustrated in Figure 80. Thermocouple T6 measured the temperature between the reactor wall and the monolith. The readings of thermocouple T6 were rather predictable for the first 16 minutes after the reaction ignition. Then the temperature suddenly dropped and fluctuated widely. During that period, the oxygen was detected by the gas probes taken immediately after monolithic catalyst suggesting that the feed breakthrough has occurred in the reactor. Hence, in spite of mitigating the gas flow bypass around the catalyst by wrapping it in a ceramic cloth, a gap between the monolith and the reactor wall opened due to the difference in the thermal-expansion of two structures.

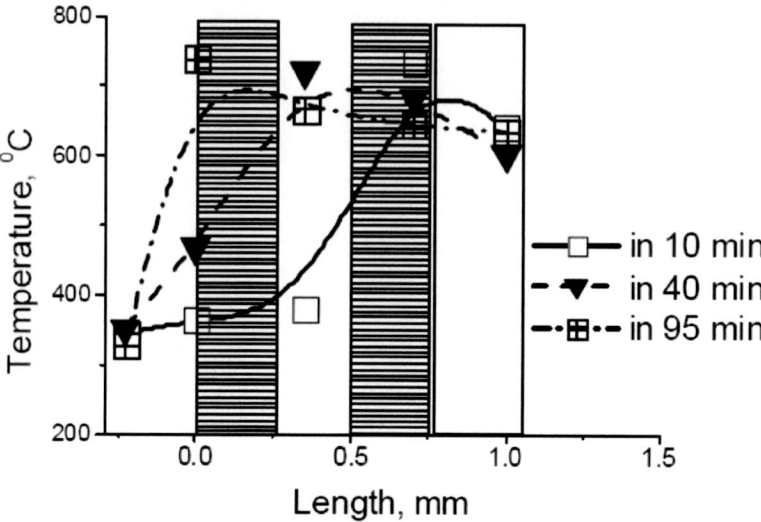

Figure 78. Temperature evolution in monolith reactor at too low superficial velocity of the gasoline-air mixture (0.26 m/s).

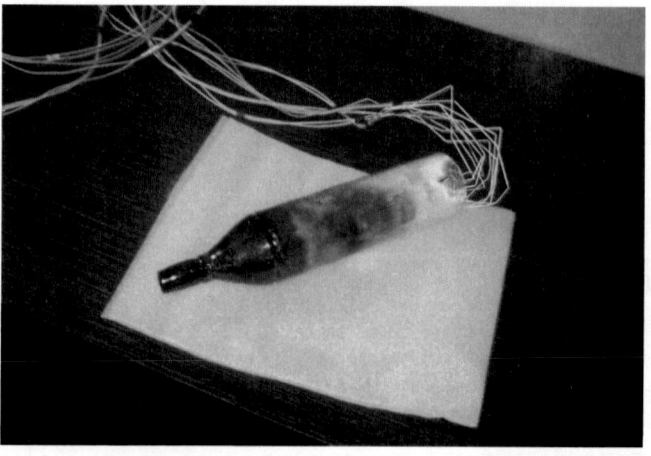

Figure 79. A view of the quartz reactor after experimental work on the diesel partial oxidation.

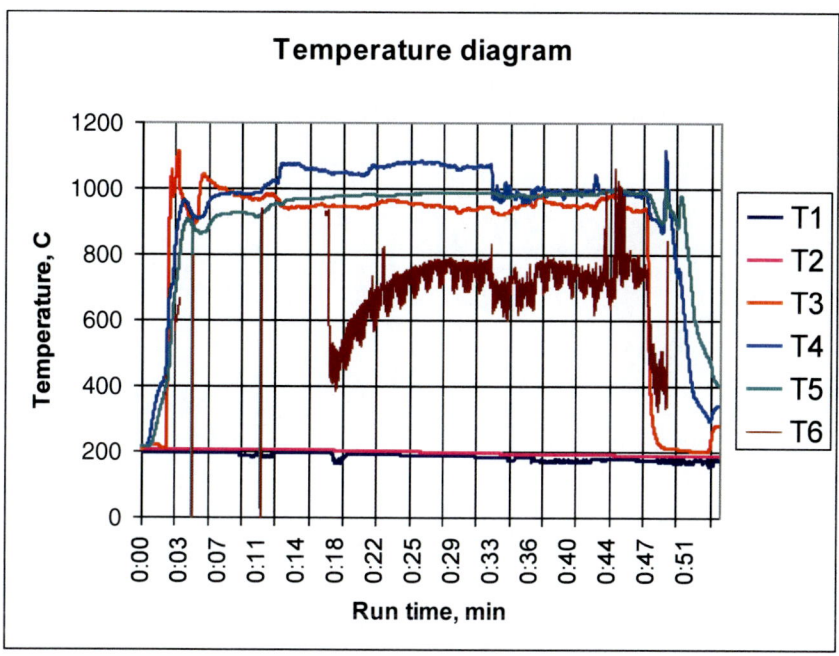

Figure 80. Temperature diagram of an experimental run with the stainless steel reactor.

It is known that the ignition of homogeneous reactions in the presence of a heterogeneous surface (the catalyst, the reactor wall, filling inert material) is characterized by a lower activation energy than that for a pure homogeneous reaction. To control such a heterogeneously initiated reactions, the reactor must be designed to account for the thermal expansion phenomena also.

CONCLUSION

Nanocomposite active components comprised of Ni (nano) particles promoted by precious metals (Pt, Ru) and supported on/diluted by fluorite-like/perovskite like oxides with a high surface/lattice oxygen mobility possess a high activity and stability to coking in the reactions of syngas generation via steam/oxysteam reforming or partial oxidation of a variety of fuels (hydrocarbons, oxygenates). By supporting these active components on monolithic substrates of different types (ceramics, refractory alloys including honeycombs, microchannel plates etc), structured catalysts are manufactures which allowed to design compact syngas generators with a high productivity and performance stability operating on a variety of fuels. The most promising systems are based on metal alloy substrates provided their functionally graded design includes a refractory dense and thin ceramic layer protecting the metal substrate from contact with aggressive high – temperature reaction stream. A high thermal conductivity of these substrates provides an efficient transfer of heat within the catalytic monolith required for the efficient conjugation of exothermal and endothermal reactions, and, thus, helping to achieve a high performance at short contact times. Mathematical modelling and detailed thermodynamic analysis of complex processes occurring in syngas generators at short contact times characterized by steep temperature and concentration gradients allowed to describe basic features of both steady-state and dynamic performance thus providing required bases for up-scaling of these devises and optimization of their performance.

ACKNOWLEDGMENTS

Support of different parts of this research by International Science and Technology Center 2529 and 3234 Projects, INTAS 05-1000005-7663 Project, European Community 6 Framework Program SOFC 600 Project, RFBR-CNRS 05-03-34761 Project in the framework of Russian French laboratory on Catalysis and Integration Project 57 of Siberian Branch of the Russian Academy of Sciences –NAN of Belarus is gratefully acknowledged.

REFERENCES

[1]   Energy and environment in the European Union: An indicator-based analysis. *Natural Resources Forum.* 2005, *29*, 360-376.

[2]   Barreto, L.; Makihira, A. and Riahi, K. *Int. J. Hydrogen Energy.* 2003, *28*, 261-360.

[3]   Rosen, M.A.; Scott, D.S. *Int. J. Hydrogen Energy.* 1998, *23*, 653–659.

[4]   Satyapal, S.; Petrovic; J., Read, C.; Thomas, G.; Ordaz, G. *Catal. Today.* 2007, *120*, 246-256.

[5]   Rostrup-Nielsen, J.R.; Sehested, J.; Nørskov, J.K. *Advances in Catalysis.* 2002, *47*, 65-139.

[6]   Wurster, R. In Fuel Cell Teach-in European Commission DGTren; L-B-Systemtechnik GmbH: Brussels, Belgium, 2002.

[7]   Liu, D-J.; Kaun, T.D.; Hsiu-Kai Liao, H.-K.; Ahmed, S. *Int. J. Hydrogen Energy.* 2004, *29*, 1035-1046.

[8]   Joensen, F.; Rostrup-Nielsen, J.R. *J. Power Sources.* 2002, *105*, 195–201.

[9]   Krumpelt, M., Krause, T.R., Carter, J.D., Kopasz, J.P.; Ahmed, S. *Catal. Today.* 2002, *17*, 3-16.

[10]   Demirbas, A. *Progress in Energy and Combustion Science.* 2007, *33*, 1–18.

[11]   Song, C. *Catal. Today.* 2002, *77*, 17–49.

[12]   Ni, M.; Leung, D. Y.C.; Leung, M. K.H.; Sumathy, K. *Fuel Processing Technology.* 2006. *87*, 461 – 472.

[13]   Navarro, R.M.; Peña, M.A.; Fierro, J.L.G. *Chem. Rev.* 2007, *107*, 3952-3991.

[14]   Bridgwater, A.V.; Czernik, S.; Piskorz, J. in: *Fast Pyrolysis of Biomass*, Bridgwater, A.V.; Eds.; CPL Press, Newbury, UK; 2002, vol. 2, pp 1–22. Czernik, S.; French, R.; Feik, C.; Chornet, E. *Ind. Eng. Chem. Res.*2002, *41*, 4209.

[15] Oasma, A.; Meier D. in: *Fast Pyrolysis of Biomass*, Bridgwater, A.V.; Eds.; CPL Press, Newbury, UK; 2002, vol. 2, pp 41–58.

[16] Takanabe, K.; Aika, K.; Seshan, K.; Lefferts, L. *J. Catal.* 2004, *227*, 101-108.

[17] Galdámez J.R.; García L.; Bilbao R. *Energy Fuels.* 2005, *19*, 1133-1142.

[18] Kechagiopoulos, P.N., Voutetakis, S.S., Lemonidou, A.A., and Vasalos, I. A., *Ind. Eng. Chem. Res.*, 2009, *48*, 1400–1408.

[19] Takanabe, K.; Aika, K.-I.; Seshan, K.; Lefferts, L. Topics Catal. 2008, *49*, 68–72.

[20] Wu, C.; Sui, M.; Yan, Y. *Chem. Eng. Technol.* 2008, *31*, 1748–1753.

[21] Choudhary, V.R.; Mamman, A.S.; Sansare, S.D. *Angew. Chem. Int. Ed. Engl.* 1992, *31*, 1189-1190.

[22] Hickman, D.A.; Schmidt, L.D. *Science.* 1993, *259*, 343-346.

[23] Hickman, D.A.; Schmidt, L.D. *J. Catal.* 1992, *138*, 267-282.

[24] Hickman, D.A.; Haupfear, E.A.; Schmidt, L.D. *Cat. Lett.* 1993, *17*, 223-237.

[25] Basini, L.; Guarinoni, A.; Aasberg-Petersen, K. *Stud. Surf. Sci. Catal.* 1998, *119*, 699-704.

[26] Bodke, A.S.; Bharadwaj, S.S.; Schmidt, L.D. *J. Catal.* 1998, *179*, 138-149.

[27] Inui, T. *Stud. Surf. Sci. Catal.* 1993, *77*, 17.

[28] Giroux, Th.; Hwang, Sh.; Liu, Y.; Ruettinger, W.; Shore, L. *Appl. Catal. B.* 2004, *55*, 179–194.

[29] Cybulski, A.; Moulijn, J.A.; Eds.; Structured Catalysts and Reactors, Chemical Industries, Vol. 71, Marcel Dekker, New York, 1998.

[30] Chen, J.; Ring, Z. Monolith Catalysts/Reactors and Their Industrial Applications, Hyfrocarbon World 2007, 56-58. (http://www.touch briefings.com/pdf/2447/chen.pdf).

[31] Italiano, G.; Espro, C.; Arena, F.; Frusteri, F.; Parmaliana, A. *Appl. Catal A*, 2009, *357*, 58-65.

[32] Schmidt, L.D. *Stud. Surf. Sci. Catal.* 2000, *130*, 61-81.

[33] Moulijn, J., Stankewicz, A., Kapteijn, F. *Chemistry for Sustainable Development.* 2003, *11*, 3-9.

[34] Ahmed, S.; Lee, S.H.; Doss, E.; Pereira, C.; Colombo, D.; Krumpelt, M. *DOE Office of Transportation Technologies*, US, October 2000.

[35] Nilsson, M.; Karatza, X.; Lindström, B.; Pettersson, L. J. *Chem. Eng. J.* 2008, *142*, 309–317.

[36] Qi, A.; Peppley, B.; Karan, K. *Fuel Processing Technology.* 2007, *88*, 3–22.

[37] Krumpelt, M.; Krause, T.R.; Carter, J.D.; Kopasz, J.P.; Ahmed, S. *Catal. Today,* 2002, *77*, 3-16.

[38] Pettersson, L.J.; Westerholm, R. *Int. J. Hydrogen Energy.* 2001, *26*, 243-264.

[39] Ming, Q.; Healey, T.; Allen, L.; Irving, P. *Catal. Today,* 2002, *77*, 51-64.

[40] Christensen, T.S. *Appl. Catal. A Gen.* 1996, *138*, 285-309.

[41] Davidian, T.; Guilhaume, N.; Iojoiu, E.E.; Provendier, H.; Mirodatos, C. *Appl. Catal. B* 2007, *73*, 116.

[42] Domine M.E.; Iojoiu E.E.; Davidian Th.; Guilhaume N.; Mirodatos C. *Catal. Today.* 2008, *133–135*, 565–573.

[43] Arpentinier, Ph.; Cavani, F.; Trifiro, F. *The technology of catalytic oxidations.* Editions TECHNIP, France, 2001.

[44] Schmidt, L.D., Huff, M. *Catal. Today.* 1994, *21*, 443–454.

[45] Huff, M.; Torniainen, P.M.; Schmidt, L.D. *Catal. Today.* 1994, *21*, 113–128.

[46] O'Connor, R.P.; Klein, E.J.; Schmidt, L.D. *Catal. Lett.* 2000, *70*, 99–107.

[47] Subramanian, R.; Panuccio, G.J.; Krummenacher, J.J.; Lee, I.C.; Schmidt, L.D. *Chem. Eng. Sci.* 2004, *59*, 5501–5507.

[48] Krummenacher, J.J.; Schmidt, L.D. *J. Catal.* 2004, *222*, 429–438.

[49] Bi, J.-L.; Hsu, S.-N.; Yeh, Ch.-T.; Wang, Ch.-B. *Catal. Today.* 2007, *129*, 330-335.

[50] Traxel, B.E.; Hohn, K.L. *Appl. Catal A* 2003, *244*, 129–40.

[51] Palm, C.; Cremer, P.; Peters, R.; Stolten, D. *J. Power Sources.* 2002, *106*, 231–237.

[52] Deluga, G.A.; Salge, J.R.; Schmidt, L.D.; Verykios, X.E. Science. 2004, *303*, 993–997.

[53] Wanat, E.C.; Suman, B.; D. Schmidt, L. *J. Catal.* 2005, *235*, 18–27.

[54] Vagia, E.Ch.; Lemonidou, A.A. *Int. J. Hydrogen Energy.* 2007, *32*, 212–223.

[55] Song, C.; Eser, S.; Schobert, H.H.; Hatcher P.G. *Energy Fuels.* 1993, *7*, 234-243.

[56] Song, C.; Lai, W.C.; Schobert, H.H. *Ind. Eng. Chem. Res.* 1994, *33*, 534-547.

[57] Rostrup-Nielsen, J.R. in *Catalysis - Science and Technology.* Anderson, J.R.; Boudart, M., Eds.; Springer Verlag, Berlin, 1984, Vol. 5, p. 1-117.

[58] Davda, R.; Shabaker, J.; Huber, G.; Cortright, R.; Dumesic, J. *Appl. Catal B* 2005, *56*, 171–186. G. W. Huber, S. Iborra, and A. Corma, *Chem. Rev.* 2006, *106*, 4044-4098.

[59] Sutton, D.; Kelleher, B.; Ross, J. R.H. *Fuel Processing Technology.* 2001, *73*, 155–173.

[60] Slinn, M.; Kendall, K.; Mallon, Ch.; Andrews, J. *Bioresource Technology.* 2008, *99*, 5851–5858.

[61] Prins, M., Ptasinski, K.; Janssen, F. *Chem. Eng. Sci.* 2003, *58*, 1003–1011.

[62] Peña, M.A.; Cómez, J.P.; Fierro, J. L.G. *Appl. Catal. A* 1996, *144*, 7-57.

[63] Vernon, P.D.F.; Green, M.L.H.; Cheetham, A.K.; Ashcroft, A.T. *Catal. Today.* 1992, 13, 417-426.

[64] Nakagawa, K.; Ikenaga, N.; Kobayashi, T.; Suzuki, T. *Catal. Today.* 2001, *64*, 31-40.

[65] Hofstad, K. H.; Hoebink, J.H.B.; Holmen, A.; Marin, G.B. *Catal. Today.* 1998, *40*, 157-170.

[66] Schmidt, L.D. *Stud. Surf. Sci. Catal.* 2000, *130*, 61-69.

[67] Au, C.T.; Ng, C.F.; Liao, M.S. *J. Catal.* 1999, *185*, 12-22.

[68] Inui, T. *Catalysis.* 2002, *16*, 133-154.

[69] Mattos, L.V.; de Oliveira, E.R.; Resende, P.D.; Noronha, F.B.; Passos, F.B. *Catal. Today.* 2002, *77*, 245-256.

[70] O'Connor, A.M.; Ross, J.R.H. *Catal. Today.* 1998, *46*, 203-210.

[71] Prettre, M.; Eichner, C.H.; Perrin, M. *Trans. Faraday Soc.* 1946, *42*, 335-340.

[72] Vermeiren, W.J.M.; Blomsma, E.; Jacobs, P.A. *Catal. Today.* 1992, *13*, 427-436.

[73] Choudhary, V.R.; Rane, V.H.; Rajput, A.M. *Catal. Lett.* 1993, *22*, 289-297.

[74] Chu, W.; Yan, Q.; Liu, S.; Xiong, G. *Stud. Surf. Sci. Catal.* 2000, *130*, 3573-3578.

[75] Chu, W.; Yan, W.G.; Liu, X.; Li, Q.; Yu, Z.L.; *Stud. Surf. Sci. Catal.* 1998, *119*, 849-854.

[76] Meadowcroft, D.B. *Nature.* 1970, *226*, 847-853.

[77] Seiyama, T. *Catal. Rev.-Sci. Eng.* 1992, *34*, 281-300.

[78] Chen, P.; Zhang, H.B.; Lin, G.D.; Tsai, K.R. *Appl. Catal. A* 1998, *166*, 343-350.

[79] Flytzani-Stephanopoulos, M.; Voecks, G.E. *Int. J. Hydrogen Energy.* 1983, *8*, 539-548.

[80] Krüger, R.; Voß A. *VDI-Berichte.* 2000, *1565*, 493–528.

[81] Sadykov, V.A.; Kuznetsova, T.G.; Frolova-Borchert, Yu.V.; Alikina, G.M.; Lukashevich, A.I.; Rogov, V.A.; Muzykantov V.S.; Pinaeva, L.G.; Sadovskaya, E.M.; Ivanova, Yu.A.; Paukshtis, E.A.; Mezentseva, N.V.; Batuev, L.Ch.; Parmon, V.N.; Neophytides, S.; Kemnitz, E.; Scheurell, K.; Mirodatos, C.; van Veen, A.C. *Catal. Today.* 2006, *117*, 475-483.

[82] Sadykov, V. A., Kuznetsova, T. G., Alikina, G. M., Frolova, Yu. V., Lukashevich, A. I., Muzykantov, V. S., Rogov, V. A., Batuev, L.Ch.,

Kriventsov,V. V., Kochubei,D. I., Moroz, E. M., Zyuzin, D. A., Paukshtis, E. A., Burgina, E. B., Trukhan, S.N., Ivanov, V.P., Pinaeva, L.G., Ivanova,Yu. A., Kostrovskii, V.G., Neophytides, S., Kemnitz, E., Scheurel, K., Mirodatos, C., In: New Topics in Catalysis Research (Ed. D.K. McReynolds), Nova Science Publishers Inc., NY, USA, 2006, p. 97-196.

[83] Pavlova, S.N.; Sazonova, N.N.; Ivanova, J.A.; Sadykov, V.A.; Snegurenko, O.I.; Rogov, V.A.; Zolotarskii, I.A.; Moroz, E.M. *Catal. Today.* 2004, *91-92*, 299-303.

[84] Parmon, V.N.; Kuvshinov, G.G.; Sadykov, V.A.; Sobyanin, V.A. *Stud. Surf. Sci. Catal.* 1998; *119*, 677-684.

[85] Pavlova, S.N.; Sadykov,, V. A.; Paukshtis, E. A.; Burgina, E. B.; Degtyarev, S. P.;Kochubei, D. I.; Saputina, N. F.; Kalinkin, A. V.; Maximovskaya, R. I.; Zaikovskii, V. I.; Roy, R.; Agrawal, D. *Stud. Surf. Sci. Catal.* 1998, *119*, 759-764.

[86] Simakov, A.V.; Pavlova, S.N.; Sazonova, N.N.; Sadykov, V.A.; Snegurenko, O.I.; Rogov, V.A.; Parmon, V.N.; Zolotarskii, I.A.; Kuzmin, V.A.; Moroz, E.M. *Chemistry for Sustainable Development.* 2003, *11*, 263-270.

[87] Roh, H.-S.; Dong, W.-Sh.; Jun, K.-W.; Park, S.-E. *Chem. Lett.* 2001, 88-89.

[88] Dong, W.-Sh.; Jun, K.-W.; Roh, H.-S.; Liu, Zh.-W.; Park, S.-E. *Catal. Lett.* 2002, *78*, 215-222.

[89] Roh, H.-S.; Jun, K.-W.; Dong, W.-Sh.; Park, S.-E.; Baek, Y.-S. *Catal. Lett.* 2001, *74*, 31-36.

[90] Takeguchi, T.; Furukawa, Sh.-N.; Inoue, M. *J. Catal.* 2001, *202*, 14-24.

[91] Takeguchi, T.; Furukawa, Sh.-N.; Inoue, M.; Eguchi, K. *Appl. Catal. A.* 2003, *240*, 223-233.

[92] Roh, H.-S.; Jun, K.-W.; Park, S.-E. *Appl. Catal. A.* 2003, *251*, 275-283.

[93] Dong, W.-Sh.; Roh, H.-S.; Jun, K.-W.; Park, S.-E.; Oh, Y.-S. *Appl. Catal. A.* 2002, *226*, 63-72.

[94] Lee, S.-H.; Cho, W.; Ju, W.-S.; Cho, B.-H.; Lee, Y.-Ch.; Baek, Y.-S. *Catal. Today.* 2003, *87*, 133-137.

[95] Wei, J.-M.; Xu, B.-Q.; Li, J.-L.; Cheng, Zh.-X.; Zhu, Q.-M. *Appl. Catal. A.* 2000, *196*, L167-L172.

[96] Roh, H.-S.; Jun, K.-W.; Baek, Y.-S.; Park, S.-E. *Catal. Lett.* 2002, *81*, 147-151.

[97] Wang, J.B.; Kuo, L.-E.; Huang, T.-J. *Appl. Catal. A.* 2003, *249*, 93-105.

[98] Asami, K.; Li, X.; Fujimoto, K.; Koyama, Y.; Sakurama, A.; Kometani, N.; Yonezawa, Y. *Catal. Today.* 2003, *84*, 27-31.

[99]   Xu, B.-Q.; Wei, J.-M.; Yu, Y.-T.; Li, J.-L.; Zhu, Q.-M. *Topics in Catalysis.* 2003, *22*, 77-85.

[100] Verykios, X.E. *Int. J. Hydrogen Energy.* 2003, *28*, 1045-1063.

[101] Wang, H.Y.; Ruckenstein, E. *Appl. Catal. A.* 2000, *204*, 143-152.

[102] Bernal, S.; Calvino, J.J.; Gatica, J.M.; Cartes, C.L.; Pintado, J.M.; In *Catalysis by Ceria and Related Materials;* Trovarelli, A.; Ed.; Catalytic Science Series; Imperial College Press: London, UK, 2002. Vol. 2, pp 85-168.

[103] Kašpar, J.; Fornasiero, P. In *Catalysis by Ceria and Related Materials;* Trovarelli, A.; Ed.; Catalytic Science Series; Imperial College Press: London, UK, 2002. Vol. 2, pp.217-241.

[104] Pechini M.P. US Patent 1967, 3 330 697.

[105] Wei, J.; Iglesia, E. *J. Phys. Chem.* 2004, *108*, 4094-4103.

[106] Kašpar, J.; Fornasiero, P.; Graziani, M. *Catal. Today.*1999, *50*, 285-298.

[107] Laosiripojana, N.; Chadwick, D.; Assabumrungrat, S. *Chem. Eng. J.* 2008, *138*, 264–273.

[108] Mogensen, M., Sammes N.M.; Tompsett, G.A. *Solid State Ionics.* 2000, *129*, 63-94.

[109] Fornasiero, P.; Dimonte, R.; Rao, G. R.; Kaspar, J.; Meriani, S.; Trovarelli, A.; Graziani, M. *J. Catal.* 1995, *151*, 168-177.

[110] Gubanova, E. L.; Van Veen, A.; Mirodatos, C.; Sadykov, V. A.; Sazonova, N. N. *Russian J. General Chem.* 2008, *78*, 2191–2202.

[111] Gubanova, E. L., Etude expérimentale et modélisation de l'oxydation partielle du méthane en gaz de synthèse sur réacteur catalytique monolithique à temps court, PhD Thesis, l'Université Claude Bernard Lyon1, 2008.

[112] Hu, Y.H.; Ruckenstein, E. *J. Phys. Chem. A.* 1998, *102*, 10568–10571.

[113] Sadykov, V.A.; Kriventsov, V.V.; Moroz, E.M.; Borchert,Yu.V.; Zyuzin, D.A.; Kol'ko, V.P.; Kuznetsova, T.G.; Ivanov, V.P.; Boronin, A.I.; Mezentseva, N.V.; Burgina, E.B.; Ross, J. *Solid State Phenomena.* 2007, 128, 81-88.

[114] Sadovskaya, E. M.; Ivanova, Yu. A.; Pinaeva, L. G.; Grasso, G.; Kuznetsova, T. G.; Van Veen, A. ; Sadykov, V. A.; and Mirodatos, C. *J. Phys. Chem. A* 2007, *111*, 4498-4505

[115] Sadykov, V.A.; Kuznetsova, T.G.; Alikina, G.M.; Frolova, Y.V.; Lukashevich, A.I;.. Potapova, Y.V; Muzykantov, V.S.; Rogov, V.A.; Kriventsov, V.V.; Kochubei, D.I.; Moroz, E.M.; Zyuzin, D.I.; Zaikovskii, V.I.; Kolomiichuk, V.N.; Paukshtis, E.A.; Burgina, E.B.; Zyryanov, V.V.;

Uvarov, N.F.; Neophytides, S.; Kemnitz, E. *Catal. Today.* 2004, *93-95*, 45-53.

[116] Mirodatos, C.; Schuurman, Y.; van Veen, A.C.; Sadykov, V.A.; Pinaeva L.G.; Sadovskaya, E.M. *Stud. Surf. Sci. Catal.* 2007, *167*, 287-292.

[117] Pavlova, S.; Tikhov, S.; Sadykov, V.; Dyatlova, Y.; Snegurenko, O.; Rogov, V.; Vostrikov, Z.; Zolotarskii, I.; Kuzmin V.; Tsybulya, S. *Stud. Surf. Sci. Catal.* 2004, *147*, 223-228.

[118] Tikhov, S.F.; Usoltsev, V.V.; Sadykov, V.A.; Pavlova, S.N.; Snegurenko, O.I.; Gogin, L.L.; Vostrikov, Z.Yu.; Salanov, A.N.; Tsybulya, S.V.; Litvak, G.S.; Golubkova G.V.; Lomovskii, O.I. *Stud. Surf. Sci. Catal.* 2006, *162*, 641-648.

[119] Pavlova, S.; Sazonova, N.; Sadykov, V.; Pokrovskaya, S.; Kuzmin, V.; Alikina, G.; Lukashevich A.; Gubanova, E. *Catal. Today*, 2005, *105*, 367-371.

[120] Kuznetsova, T.G.; Sadykov, V.A.; Moroz, E.M.; Trukhan, S.N.; Paukshtis, E.A.; Kolomiichuk, V.N.; Burgina, E.B.; Zaikovskii, V.I.; Fedotov, M.A.; Lunin, V.V.; Kemnitz, E. *Stud. Surf. Sci. Catal.* 2000, *143*, 659-667.

[121] Pavlova, S.N.; Sazonova, N.N.; Sadykov, V.A.; Alikina, G.M.; Lukashevich, A.I.; Gubanova E.; Bunina, R.V. *Stud. Surf. Sci. Catal*, 2007, *167*, 343-348.

[122] Sadykov, V.; Pavlova, S.; Vostrikov, Z.; Sazonova, N.; Gubanova, E.; Bunina, R.; Alikina, G.; Lukashevich, A.; Pinaeva, L.; Gogin, L.; Pokrovskaya, S.; Skomorokhov, V.; Shigarov, A.; Mirodatos, C.; van Veen, A.; Khristolyubov A.; Ulyanitskii, V. *Stud. Surf. Sci. Catal.* 2007, *167*, 361-366.

[123] Sadykov, V.A.; Pavlova, S. N.; Bunina, R.V.; Alikina, G.M.; Tikhov, S.F.; Kuznetsova, T.G.; Frolova, Yu.V.; Lukashevich, A.I.; Snegurenko, O.I.; Sazonova, N.N.; Kazantseva, E.V.; Dyatlova, Yu.N.; Usol'tsev, V.V.; Zolotarskii, I.A.; Bobrova, L.N.; Kuz'min, V.A.; Gogin, L.L.; Vostrikov, Z.Yu.; Potapova, Yu.V.; Muzykantov, V.S.; Paukshtis, E.A.; Burgina, E.B.; Rogov, V.A.; Sobyanin, V.A.; Parmon, V.N. *Kinetics and Catalysis.* 2005, *46*, 227–250.

[124] Pavlova, S.N.; Tikhov, S.F.; Sadykov, V.A.; Snegurenko, O.I.; Dyatlova, Yu.N.; Zolotarskii, I.A.; Kuzmin, V.A.; Bobrova, L.N.; Vostrikov Z.Yu. *Patents RF.* 2005, 2244589, 2248240.

[125] van Rossum, G.; Kersten, S.R.A.; van Swaaij, W.P.M. *Ind. Eng. Chem. Res.* 2007, *46*, 3959–3967.

[126] Rioche, C.; Kulkarni, S.; Meunier, F.C.; Breen, J.P.; Burch, R. *Appl. Catal. B: Envir.* 2005, *61*, 130–139.

[127] Tomishige, K.; Asadullah, M. In *Progress in Catalysis Research*, Bevy, L. P., Ed.; 2005, Nova Science Publishers, Inc. NY, USA, pp 1-39.

[128] Takanabe, K.; Aika, K.; Seshan, K.; Lefferts, L. *Chem. Eng. J.* 2006, *120*, 133–137.

[129] Fatsikostas, A.N.; Kondarides, D.I.; Verykios, X.E. *Catal. Today.* 2002, *75*, 145-155.

[130] Jamsak, W.; Assabumrungrat, S.; Douglas P.L.; Laosiripojana, N.; Suwanwarangkul, R.; Charojrochkul, S.; Croiset, E. *Chem. Eng. J.* 2007, *133*, 187-194.

[131] Haryanto, A.; Fernando, S.; Murali, N.; Adhikari, S. *Energy Fuels* 2005, *19*, 2098-2106.

[132] Breen, P.; Burch, R.; Coleman, H.M. *Appl. Catal. B* 2002, *39*, 65-74.

[133] Basagiannis, A.C.; Verykios, X.E. *Appl. Catal B* 2008, *82*, 77–88.

[134] Fierro, V.; Klouz, V.; Akdim, O.; Mirodatos, C. *Catal. Today.* 2002, *75*, 141-144.

[135] Fierro, V.; Akdim, O.; Mirodatos, C. *Green Chem.* 2003, *5*, 20-24.

[136] Fierro, V.; Akdim, O.; Provendier, H.; Mirodatos, C. *J. Power Sources.* 2005, *145*, 659-666.

[137] Frusteri, F.; Freni, S.; Chiodo, V.; Donato, S.; Bonura, G.; Cavallaro, S. *Int. J. Hydrog. Energy.* 2006, *31*, 2193-2199.

[138] Kugai, J.; Subramani, V.; Song, C.; Engelhard, M.H.; Chin, Y. *J. Catal.* 2006, *238*, 430-440.

[139] Cai, W.; Zhang, B.; Li, Y.; Xu, Y.; Shen, W. *Catal. Commun.* 2007, *8*, 1588-1594.

[140] Cavallaro, S.; Chiodo, V.; Vita, A.; Freni, S. J. *Power Sources.* 2003, *123*, 10-16.

[141] Cavallaro, S.; Mondello, N.; Freni, S. J. *Power Sources.* 2001, *102*, 198-204.

[142] Liguras, D.K., Goundani, K.; Verykios, X.E. *Int. J. Hydrogen Energy.* 2004, *29*, 419427.

[143] Zhang, B.; Tang, X.; Li, Y.; Cai, W.; Xu, Y.; Shen, W. *Catal.Commun.* 2006, *7*, 367-372.

[144] Cai, W.; Wang, F.; Zhan, E.; Van Veen, A.C.; Mirodatos, C.; Shen, W. *J. Catal.* 2008, *257*, 96–107

[145] Zhang, B.; Cai, W.; Li, Y.; Xu, Y.; Shen, W. *Int. J. Hydrogen Energy.* 2008, *33*, 4377-4386.

[146] Degenstein, N.J.; Subramanian, R.; Schmidt, L.D. *Appl. Catal. A.* 2006, 305, 146-159.

[147] Nguyen, B.N.T.; Leclerc, C.A. *Int. J. Hydrogen Energy.* 2008, *33*, 1295-1303.

[148] Matas Guell, B.; Babich, I.; Seshan, K.; Lefferts, L. *J. Catal.* 2008, *257*, 229–231.

[149] Takanabe, K.; Aika, K.; Inazu, K.; Baba, T.; Seshan, K.; Lefferts, L. *J. Catal.* 2006, *243*, 263-269.

[150] Aupretre, F.; Descorme, C.; Duprez, D.; Casanave, D.; Uzio, D. *J. Catal.* 2005, *233*, 464-477.

[151] Wanat, E.C.; Venkataraman, K.; Schmidt, L.D. *Appl. Catal. A.* 2004, *276*, 155-162.

[152] Diagne, C.; Idriss, H.; Kiennemann, A. *Catal. Commun.* 2002, *3*, 565-571.

[153] Salge, J.R.; Deluga, G.A.; Schmidt, L.D. *J. Catal.* 2005, *235*, 69-78.

[154] Vagia, E. Ch.; Lemonidou, A. A. *Appl. Catal. A: Gen.* 2008, *351*, 111-121.

[155] . Slinn, M.; Kendall, K. *Bioresource Technology* 2009, *100, 2324-2327.*

[156] Henderson, M.A. *Surface Science Reports.* 2002, *46*, 1–308.

[157] Sadykov, V.A.; Mezentseva, N.V.; Bunina, R.V.; Alikina, G.M.; Lukashevich, A.I.; Kharlamova, T.S.; Rogov, V.A.; Zaikovskii, V.I.; Ishchenko, A.V.; Krieger, T.A.; Bobrenok, O.F.; Smirnova, A.; Irvine, J.; Vasylyev, O.D. *Catal. Today.* 2008, *131*, 226-237.

[158] Sadykov, V.A.; Mezentseva, N.V.; Bunina, R.V. *Mater. Res. Soc. Symp. Proc.* 2007, *972, AA03-06.*

[159] Sadykov, V.; Mezentseva, N.; Alikina, G.; Lukashevich, A.; Muzykantov, V.; Bunina, R.; Boronin, A.; Pazhetnov, E.; Paukshtis, E.; Kriventsov, V.; Smirnova, A. A.; Vasylyev, O. A.; Irvine, J. A.; Bobrenok, O.; Voronin, V.; Berger, I. *Mater. Res. Soc. Symp. Proc.* 2007, *1023*, JJ02-07.

[160] Sadykov, V.A.; Mezentseva, N.V.; Bunina, R.V.; Alikina, G.M.; Lukashevich, A.I.; Borchert, Yu.V.; Kuznetsova, T.G.; Ivanov, V.P.; Trukhan, S.N.; Paukshtis, E.A.; Muzykantov, V.S.; Kuznetsov, V.L.; Rogov, V.A.; Ross, J.R.H.; Kemnitz, E.; Sheurell, K. *Solid State Phenom.* 2007, *128*, 239-248.

[161] Sadykov, V.; Mezentseva, N.; Alikina, G.; Lukashevich, A.; Muzykantov, V. ; Kuznetsova, T.; Batuev, L.; Fedotov, M.; Moroz, E.; Zyuzin, D.; Kolko, V.;Kriventsov, V.; Ivanov, V. ; Boronin, A.; Pazhetnov, E.; Zaikovskii, V.; Ishchenko, A.; Rogov, V.; Ross, J.; Kemnitz, E. *Mater. Res. Soc. Symp. Proc.* 2007, *988*, QQ04-06.

[162] Grasso, G.; Harji, B.; Xue, E.; Belochapkine, S.; Ross, J.R.H. *Catal. Today.* 2003, *81*, 369-375.

[163] Sadykov, V.; Mezentseva, N.; Alikina, G.; Bunina, R.; Rogov, V.; Krieger, T.; Belochapkine, S.; Ross, J.R.H. *Catal.Today.* 2009, *145*, 127-131.

[164] Yaseneva, P.; Pavlova, S.; Sadykov, V.; Alikina, G.; Lukashevich, A.; Rogov, V.; Belochapkine, S.; Ross, J.R.H. *Catal. Today*. 2008, *137*, 23-28.

[165] Topsøe hydrogen package plants, www.topsoe.com.

[166] Agrell, J.; Lindstrom, B.; Pettersson, L.J.; Jaras, S.G. in: *Catalysis—Specialist Periodical Reports. Spivey J.J.; Ed.*, 2002, *16*, 67-132.

[167] Harold, M.P.; Nair, B.; Kolios, G. *Chem. Eng. Sci*. 2003, *58*, 2551-2571.

[168] Breen, J.P.; Ross, J.R.H. *Catal. Today*. 1999, *51*, 521-533.

[169] Clancy, P.; Breen, J. P.; Ross, J.R.H. *Catal. Today*. 2007, *127*, 291-294.

[170] Mastalir, A.; Frank, B.; Szizybalski, A.; Soerijanto, H.; Deshpande, A.; Niederberger, M.; Schomäcker, R.; Schlögl, R;. Ressler, T. *J. Catal*. 2005, *230*, 464-475.

[171] Yaseneva, P.; Pavlova, S.; Sadykov, V.; Moroz, E.; Burgina, E.; Dovlitova, L.; Rogov, V.; Badmaev, S.; Belochapkine, S.; Ross, J. R.H. *Catal. Today*. 2008, *138*, 175-182.

[172] Iordanidis, A.A.; Kechagiopoulos, P.N.; Voutetakis, S.S.; Lemonidou, A.A.; Vasalos, I.A. *Int. J. Hydrogen Energy*. 2006, *31*, 1058-1065.

[173] Scott, D. S.; Piskorz, J.; Radhlein, D. *Ind. Eng. Chem. Process Res. Dev*. 1985, *24*, 581-590.

[174] Pinkwart, K.; Bayha, T.; Lutter, W.; Krausa,. M. *J. Power Sources*. 2004, *136*, 211-214.

[175] Ramasamy, K.K.; T-Raissi, A. *Catal. Today*. 2007, *129*, 365-371.

[176] Dietz, A.G.; Schmidt, L. D. *Catal. Lett*. 1995, *33*, 15-29.

[177] Heitnes, K.; Lindberg, S.; Rokstad, O.A.; Holmen, A. *Catal. Today*. 1994, *21*, 471-480.

[178] Heitnes, K.; Lindberg, S.; Rokstad, O.A.; Holmen, A. *Catal. Today*. 1995, *24*, 211-216.

[179] Heitnes, K.; Sperle, T.; Rokstad, O.A.; Holmen, A. *Catal. Lett*. 1997, *45*, 97-105.

[180] Groppi, G.; Airoldi, G.; Cristiani, C.; Tronconi, E. *Catal. Today*. 2000, *60*, 57-62.

[181] Groppi, G.; Tronconi, E. *Chem .Eng. Sc*. 2000, *55*, 2161-2171.

[182] Groppi, G.; Tronconi, E. *Chem. Eng. Sci*. 2000, *55*, 6021-6036.

[183] Kucharczyk, B.; Tylus W.; Kępiński. L. *Appl. Catalysis B* 2004, *49*, 27-37.

[184] Avila, P.; Montes, M.; Miróc, E. E. *Chem. Eng. J*. 2005, *109*, 11-36.

[185] Put, J.; Vleugels, G.; Anne., O.; Van der Biest. O. *Colloids and Surfaces A:* 2003, *222*, 223-232.

[186] Tikhov, S.F.; Romanenkov, V.E.; Sadykov, V.A.; Parmon, V.N. and Rat'ko, A.I. *Kinet. Catal*. 2005, *46*, 641-659.

[187] Tikhov, S.F.; Sadykov, V.A.; Salanov, A.N.; Potapova, Yu. V.; Tsybulya, S.V.; Litvak, G.S.; Pavlova S.N. *Mater. Res. Soc. Symp.*. 1998 , *497*, 200-206.

[188] Sadykov, V.; Pavlova, S.; Snegurenko, O.; Vostrikov, Z.; Tikhov., S.; Kuzmin, V.; Parmon, V.; Ulianitsky, V.; Brizitsky, O.; Khristolubov, A.; Terentiev, V. *Stud. Surf. Sci. Catal.* 2007, *172*, 241-244.

[189] Bobrova, L.; Zolotarskii, I.; Sadykov, V.; Pavlova, S.; Snegurenko, O.; Tikhov, S.; Korotkich, V.; Kuznetsova, T.; Sobyanin, V.; Parmon, V. *Chem. Eng. J.* 2005, *107*, 171-179.

[190] Bobrova, L.; Korotkich, V.; Sadykov, V.; Parmon, V. *Chem. Eng. J.* 2007, *134*, 145-152.

[191] Bobrova, L.; Zolotarskii, I.; Sadykov, V.; Sobyanin, V. *Int. J. Hydrogen Energy.* 2007, *32*, 3698-3704.

[192] Bobrova, L.; Vernikovskaya, N.; Sadykov, V. *Catal. Today.* 2009, *144*, 185-200.

[193] Rostrup-Nielsen, R. *Catal. Today.* 2002, *71*, 243-247.

[194] Docter, A.; Lamm, A. J. *Power Sources.* 1999, *84*, 194-200.

[195] Krummenacher, J. J.; West, K.N. and Schmidt, L.D. *J. Catal.*. 2003, *215*, 332-343.

[196] Pacheco, M.; Sira J. and Kopasz, J. *Appl. Catalysis A.* 2003, *250*, 161-175.

[197] Clark, A.Q.; McBsin, S.E.; Kilner, J. *Fluid Phase Equilibria.* 1997, *133*, 239-246.

[198] Crabtree, R.H. *Chem. Rev.* 1995, *95*, 987-1007.

[199] Seo, LY.-S.; Shiley, A.; Kolaczkowski, S.T. *J. Power Sources.* 2002, *108*, 213-225.

[200] Vagia, E.Ch.; Lemonidou, A.A. *Int. J. Hydrogen Energy.* 2008, *33(10)*, 2489-2500.

[201] Czernik, S.; Bridgwater, A.V. *Energy Fuels.* 2004, *18*, 590-598.

[202] Ioannides Th. *J. Power Sources.* 2001, *92*, 17-25.

[203] Aktaş, S.; Karakaya, M.; Avcı, Ah. *Int. J. Hydrogen Energy.* 2009, *34*, 1752-1759.

[204] Veser, G. and Schmidt, L.D. *A.I.Ch.E.J.* 1996, *42*, 1977-1987.

[205] Ziauddin, M.; Veser, G.; Schmidt, L.D. *Catal. Lett.* 1997, *46*, 159-167.

[206] Park, Y.K.; Vlachos, D.G. *A.I.Ch.E.J.* 1997, *43*, 2083-2095.

[207] Bui, P.-A.; Vlachos, D.G.; Westmoreland, P.R. *Stud. Surf. Sci. Catal.* 1997, *385*, L1029-L1034.

[208] Hickman, D.A.; Schmidt, L.D. *A.I.Ch.E. Journal.* 1993, *39*, 1164-1177.

[209] Frauhammer, J.; Veser, G. *Chem. Ing. Tech.* 1998, *70*, 1020-1027.

[210] Deutschmann, O.; Schmidt, R.; Behrendt, F.; Warnatz, J. In: *Twenty Sixth Symposium (International) on Combustion*. The Combustion Institute, Pittsburgh. 1996, 1747-1754.

[211] Vernikovskaya, N.; Bobrova, L.; Pinaeva, L.; Sadykov, V.; Zolotarskii, I.; Sobyanin, V.; Buyakou, I.; Kalinin, V.; Zhdanok, S. *Chem. Eng. Journal*. 2007, *134*, 180-189.

[212] Aghalayam, P.; Park, Y.K.; Fernandes, N.; Papavassiliou, V.; Mhadeshwar, A.B.; Vlachos, D.G. *J. Catal.* 2003, *213*, 23-38.

[213] Schwiedernoch, G.R.; Tischer, S.; Correa, C.; Deutschmann. O. *Chem. Eng. Sci.* 2003, *58*, 633-642.

[214] Veser, G.; Frauhammer, J. *Chem. Eng. Sci.* 2000, *55*, 2271-2286.

[215] Basini, L.; Guarinoni, A.; Aasberg-Petersen K. *Stud. Surf. Sci. Catal.* 1998, *119*, 699-704.

[216] Basini, L.; Aasberg-Petersen, K.; Guarinoni, A.; Østberg, M. *Catal. Today.* 2001, *64*, 9-20.

[217] Somorjai, G.A. *Surf. Sci.* 1995, *335*, 10-22.

[218] Berger, R.J.; Marin, G.B. *Ind. Eng. Chem. Res.* 1999, *38*, 2582-2592.

[219] Chen, Q.; Couwenberg, P.M.; Marin, G.B. *AIChE* Journal. 1994, *40*, 521-534.

[220] Viljoen, H. J.; van Rensburg, N. F. J. *AIChE Journal.* 1995, 41, 1344-1345.

[221] Cominos, V.; Gavriilidis, A. *Chem. Eng. Sci.* 2001, *56*, 3455-3468.

[222] Khanaev, V.M.; Borisova, E.S.; Noskov, A.S. *Chem. Eng. Sci.* 2005, *60*, 5792-5802.

[223] Beretta, A.; Groppi, G.; Majocchi, L.; Forzatti, P. *Appl. Catal. A* 1999, *187*, 49-60.

[224] Bradford, M.C.J.; Vannice, M.A. *Catal. Rev.-Sci. Eng.* 1999, *41*, 1-42.

[225] Choudhary, T.V. Goodman, D.W. *J. Mol. Catal. A:* 2000, *163*, 9-18.

[226] Leites, I.L.; Sama, D.A.; Lior N. *Energy*, 2003, *28*, 55-97.

[227] Li, C.H. In: *Ullmann's Encyclopedia of Industrial Chemistry,* Copyright © 2007 by Wiley-VCH Verlag GmbH and Co. KGaA

[228] Pavlova, S.N.; Tikhov, S.F.; Sadykov, V.A.; Snegurenko, O.I.; Ulyanitskii, V. Yu.; Kuznetsova, T.G.; Zolotarskii, I.A.; Kuzmin, V.A.; Vostrikov, Z.Yu.; Bobrova, . L.N.; Kirillov, V.A.; Parmon, V.N.; Sobyanin, V.A. Patent RF. 2248932.

[229] Ulyanitskii, V. Yu.; Shterzer, A.A.; Zlobin, S.B.; Matrenin, V.I.; Schipanov, I.V.; Serykh, S. Yu.; Stikhin, A.S.; Tretyakova, L.M.; Sadykov, V.A.; Pavlova, S.N.; Tikhov, S.F.; Kuzmin, V.A. *Alternative Energetics and Ecology.* 2006, *9*, 137-144.

[230] Kaplan, I.R.; Shan-Tan Lu, H.M.; Alimi, J. *Environmental Forensics*, 2001, *2*, 231-248.

[231] Pinkwart, K.; Bayha, T.; Lutter, W.; Krausa, M. *J. Power Sources*. 2004, *136*, 211-214.

[232] Ramasamy, K.K. ; T-Raissi, A. *Catal. Today*. 2007, *129*, 365-371.

[233] Lin, K.-S.; Chowdhury, S.; Shen, C.-Ch.; Yeh, Ch.-T. *Catal. Today*. 2008, *136*, 281-290.

[234] Makarshin, L.L.; Parmon, V.N. *Russ. Chem. J.* 2006, *1*, 17- 27.

[235] Makarshin, L.L.; Andreev, D.V.; Pavlova, S.N.; Sadykov, V.A.; Snegurenko, O.I.; Privezentsev, V.V.; Gulevich, A.V.; Ulianitsky, V.Yu.; Sobyanin, V.A.; Parmon, V.N. *XVII Int. Conf. on Chemical Reactors* «CHEMREACTOR-17», Athens-Crete, Greece, May 15-19, 2006, pp 451-454.

[236] Gulevich, A.V.; Desyatov, A.V.; Zlotzovsii, A.M.; Izvolskii, I.M.; Makarshin, L.L.; Mamontov, Yu.N.; Privesentzev, V.V.; Sadykov, V.A.; Parmon, V.N. In: *Advanced Energetic Technologies on Earth and in Space*, (A. Koroteev, Ed.), Moscow, Svetlitsa, 2008.

[237] Makarshin, L.; Andreev, D.; Pavlova, S.; Sadykov, V.; Privezentsev, V.; Gulevich, A.; Sobyanin, V. *Alternative Energy and Ecology*, 2007, *2*, 132-134.

[238] Makarshin, L.L.; Andreev, D.V.; Gribovskii, A.G.; Dutov, P.M.; Khantakov, R.M.; Parmon, V.N. *Kinet. Catal.* 2007, *48*, 765–771 (Russian).

[239] Makarshin, L.; Andreev, D.V.; Gribovskiy, A.G.; Parmon, V.N. *Int. J. Hydrogen Energy*. 2007, *32*, 3864-3869.

[240] Makarshin, L.L.; Andreev, D.V.; Gribovskiy, A.G.; Hantakov, R.M.; Parmon, V.N. XVII *Int. Conf. on Chemical Reactors* „CHEMREACTOR-18", Malta, 29 September – 3 October, 2008, pp. 424-425.

INDEX